WARD VALLEY

An Examination of Seven Issues in Earth Sciences and Ecology

Committee to Review Specific Scientific and Technical
Safety Issues Related to the Ward Valley, California,
Low-Level Radioactive Waste Site

Board on Radioactive Waste Management

Commission on Geosciences, Environment, and Resources

National Research Council

NATIONAL ACADEMY PRESS
WASHINGTON D.C.   1995

Support for this study of Ward Valley was provided by the U.S. Department of the Interior, under agreement 1434-94-A-1269 (INTR/USGS).

Library of Congress Catalog Card Number 95-69191
International Standard Book Number 0-309-05288-2

Additional copies of this report are available from:
National Academy Press
2101 Constitution Avenue, N.W.
Box 285
Washington, D.C. 20055

Call 800-624-6242 or 202-334-3313 (in the Washington Metropolitan Area).

B562

Cover art by C. Etana Finkler© 1995. Ms. Finkler is a world-travelled water colorist and computer-graphics artist living in Takoma Park, Maryland. Through vividly-colored imagery, she portrays the diversity and commonality in everyday culture.

Printed in the United States of America

COMMITTEE TO REVIEW SPECIFIC SCIENTIFIC AND TECHNICAL SAFETY ISSUES RELATED TO THE WARD VALLEY, CALIFORNIA, LOW-LEVEL RADIOACTIVE WASTE SITE

GEORGE A. THOMPSON, *Chairman,* Stanford University, Stanford, California
THURE E. CERLING, University of Utah, Salt Lake City
G. BRENT DALRYMPLE, Oregon State University, Corvallis
ROBERT D. HATCHER, JR., University of Tennessee, Knoxville/Oak Ridge National Laboratory
AUSTIN LONG, University of Arizona, Tucson
MARTIN D. MIFFLIN, Mifflin and Associates, Incorporated, Las Vegas, Nevada
JUNE ANN OBERDORFER, San Jose State University, San Jose, California
KATHLEEN C. PARKER, University of Georgia, Athens
DUNCAN T. PATTEN, Arizona State University, Tempe
DENNIS W. POWERS, Consulting Geologist, Canutillo, Texas
STEPHEN J. REYNOLDS, Arizona State University, Tempe
JOHN B. ROBERTSON, HydroGeoLogic, Incorporated, Herndon, Virginia
BRIDGET R. SCANLON, University of Texas, Austin
LESLIE SMITH, University of British Columbia, Vancouver, Canada
BRUCE A. TSCHANTZ, University of Tennessee, Knoxville
SCOTT TYLER, Desert Research Institute, Reno, Nevada
PETER J. WIERENGA, University of Arizona, Tucson

Staff

INA B. ALTERMAN, Study Director
CARL A. ANDERSON, BRWM Director
KEVIN D. CROWLEY, Associate Director
REBECCA BURKA, Administrative Assistant
ELIZABETH M. LANDRIGAN, Technical Assistant

The National Academy of Sciences is a private, nonprofit, self-perpetuating society of distinguished scholars engaged in scientific and engineering research, dedicated to the furtherance of science and technology and to their use for the general welfare. Upon the authority of the charter granted to it by the Congress in 1863, the Academy has a mandate that requires it to advise the federal government on scientific and technical matters. Dr. Bruce Alberts is president of the National Academy of Sciences.

The National Academy of Engineering was established in 1964, under the charter of the National Academy of Sciences, as a parallel organization of outstanding engineers. It is autonomous in its administration and in the selection of its members, sharing with the National Academy of Sciences the responsibility for advising the federal government. The National Academy of Engineering also sponsors engineering programs aimed at meeting national needs, encourages education and research, and recognizes the superior achievements of engineers. Dr. Robert M. White is president of the National Academy of Engineering.

The Institute of Medicine was established in 1970 by the National Academy of Sciences to secure the services of eminent members of appropriate professions in the examination of policy matters pertaining to the health of the public. The Institute acts under the responsibility given to the National Academy of Sciences by its congressional charter to be an adviser to the federal government and, upon its own initiative, to identify issues of medical care, research, and education. Dr. Kenneth Shine is president of the Institute of Medicine.

The National Research Council was organized by the National Academy of Sciences in 1916 to associate the broad community of science and technology with the Academy's purposes of furthering knowledge and of advising the federal government. Functioning in accordance with general policies determined by the Academy, the Council has become the principal operating agency of both the National Academy of Sciences and the National Academy of Engineering in providing services to the government, the public, and the scientific and engineering communities. The Council is administered jointly by both Academies and the Institute of Medicine. Dr. Bruce Alberts and Dr. Robert M. White are chairman and vice chairman, respectively, of the National Research Council.

TABLE OF CONTENTS

APPENDICES

FIGURES AND TABLES

EXECUTIVE SUMMARY

INTRODUCTION

The Low-Level Radioactive Waste Policy Act (1980, amended 1985) legislates state responsibility for non-government low-level radioactive waste generated within states. California, Arizona, North Dakota, and South Dakota formed the Southwest Compact to share a disposal facility for these wastes. The Ward Valley site west of Needles, California, was investigated and proposed for the first facility to serve the compact.

As the siting and licensing process was ending, the U. S. Department of the Interior (DOI) was asked by the State of California to transfer the site lands, presently held by the Bureau of Land Management, an agency of DOI, to California for site development. While DOI was considering the land transfer, three geologists from the U. S. Geological Survey (USGS) expressed seven concerns about the site and its evaluation in a memorandum to the Secretary of the Interior, Bruce Babbitt. Although Howard Wilshire, Keith Howard, and David Miller (referred to as the Wilshire group in this report) acted as individuals rather than in official USGS capacities, the DOI asked the National Research Council (NRC) to convene a committee to evaluate their seven technical concerns prior to the DOI decision on the land transfer.

The seven issues, as originally stated in the Wilshire group's memorandum, are:

1. Potential infiltration of the repository trenches by shallow subsurface water flow.[1]
2. Transfer of contaminants through the unsaturated zone and potential for contamination of ground water.
3. Potential for hydrologic connection between the site and the Colorado River.
4. No plans are revealed for monitoring ground water or the unsaturated zone downgradient from the site.
5. Engineered flood control devices like those proposed have failed in past decades at numerous locations across the Mojave Desert.
6. Alluvium and colluvium derived from Cretaceous granite appears to make a very high quality tortoise habitat. Sacrifice of such habitat cannot be physically compensated.
7. Misconceptions about revegetation enhancement may interfere with successful reestablishment of the native community.

The committee's charge (see Appendix A of this report for details) was to evaluate the validity of these seven issues.

---

[1] This refers to subsurface *lateral* flow as confirmed by the Wilshire group.

It should be noted that the committee was not asked to and did not take any position on the overall suitability of the Ward Valley site for a LLRW disposal facility. Although the seven concerns raised by the Wilshire group relate to site suitability, this evaluation of the technical validity of these concerns does not constitute approval or disapproval of the site.

MAJOR CONCLUSIONS REGARDING THE SEVEN ISSUES

The committee offers the following summary of the major conclusions to the seven technical issues related to the Ward Valley site that it was asked to review, with a cautionary note: first, as noted in the footnote on page 1 of Chapter 3, Issues 1 and 2 of the Wilshire group have been reversed in order, so that the committee's Issue 1, the potential for transfer of contaminants through the unsaturated zone, is the Wilshire group's Issue 2 and vice-versa; secondly, *these conclusions should not be read without, nor taken out of context of, the discussions that describe the bases for the conclusions, the limitations of the data, and the levels of uncertainty which may accompany some of these conclusions.* These are summarized briefly in this Executive Summary and extensively discussed in the body of the report.

****

ISSUE 1 (Issue 2 of the Wilshire group): GENERAL CONCLUSION: The committee concludes from multiple lines of evidence that the unsaturated zone at the Ward Valley site is very dry, and that recharge or potential transfer of contaminants through the unsaturated zone to the water table, as proposed by the Wilshire group, is highly unlikely. However, because of the limitations of the data, the committee recommends specific initial baseline and subsequent monitoring measurements, summarized on page 10 of this Executive Summary, to enhance the data base for monitoring the complex unsaturated zone.[2]

Discussion of Issue 1

Issue 1 Subissues

The Wilshire group divided the evaluation in the license application of the nature of water movement in the unsaturated zone into five subissues that dealt with (1) the adequacy of the treatment of the unsaturated zone variability and complexity; (2) the possibility of rapid water migration down preferential pathways; (3) the tritium measurements at 30 meter depth

---

[2] Two committee members, J. Oberdorfer and M. Mifflin, dissented from this conclusion. Their statements can be found in Appendices E and F at the end of this report.

suggesting rapid vertical water transport; (4) possible recharge to the ground water below the major drainage, Homer Wash; and (5) a possible interpretation of stable and radioactive isotopes in the ground water suggesting recent recharge to the ground water.

- **With respect to the subissues, two unresolved data sets remain: the observed vertical hydraulic gradient between two monitoring wells, relating to subissue (2) and the presence of tritium in the unsaturated zone (subissue (3)).**[3] **The majority of the committee considers that subissue (1), unsaturated zone variability and complexity, has been adequately addressed in the modeling and analyses of the unsaturated zone, with the exception of the modeling of a complete cover failure; subissue (2), the presence of preferential pathways, is not supported by any consistent evidence for rapid downward water migration or ground water recharge below the site, despite arguments to the contrary[3], as discussed later in this Executive Summary; subissue (4), recharge below Homer Wash, is likely but will have no consequences for the containment of contaminants because of the distance of the wash from, and its elevation below, the site and the waste trenches as presently designed; and subissue (5), recent recharge to the ground water below the site, is not supported by the solute concentrations in the ground-water chemistry.**

With respect to subissue (2), monitoring wells WV-MW-01 and WV-MW-02 show an apparent downward hydraulic gradient that could be caused by local recharge, an explanation that is inconsistent with most other data. Any deviations from the vertical in the boreholes could lead to erroneous depth measurements because the measurement of depth would be the distance down the borehole to the water table rather than the actual depth.

- **The committee finds that the cause(s) of the observed vertical gradient in the saturated zone cannot be conclusively determined with the available data. The most probable sources of the apparent gradient are measurement and drilling errors.** Detailed discussion of this issue and conclusion can be found in Chapter 3 in the section on ground water gradients.
- **Regarding subissue (3), the committee finds that the conclusion in the license application that gas diffusion is responsible for the tritium reported in the unsaturated zone is conceptually incorrect. The committee concludes that inappropriate sampling procedures most probably introduced atmospheric tritium into the samples. Except for three data points at depths of 5.1 m and 5.4 m, the tritium data from deeper levels (11-30 m) are not distinguishable from zero owing to inadequate evaluation of the uncertainty of the tritium values resulting from the sample-collection procedure. The three results from the uppermost sampling depths may represent atmospheric contamination, or they may indicate small amounts of shallow infiltration. Due to these uncertainties, the tritium data are not adequate for the evaluation of infiltration. The**

---

[3] The dissenting opinions of Appendices E and F include dissent from the committee's conclusions on these subissues.

committee recommends that further analysis and sampling be conducted to resolve this subissue. Detailed discussion of this issue is in Chapter 3 in the section on environmental tracers and in the conclusions.

Basis for General Conclusions

The committee reviewed multiple lines of evidence to evaluate water flux in the unsaturated zone at the Ward Valley site. We based our conclusion that the unsaturated zone is very dry in part on the following information: (1) In 82 samples from near the surface to 27 m depth, water contents were generally very low (94 percent of the samples had water contents less than 10 percent, and 6 percent of the samples had water contents between 10 and 15 percent); (2) Water content monitored in a neutron probe access tube installed to 6-m depth showed that the maximum depth of penetration of water after rainfall was about one meter; (3) Water potentials monitored to 30-m depth were very low (-3 to -6 MPa); (4) Chloride concentrations measured in three boreholes to 30-m depth were very high (up to 15g/l), suggesting little infiltration and downward percolation of water since accumulation of the chloride. The time required to accumulate these large quantities of chloride to 30-m depth was calculated to be approximately 50,000 yr; (5) Estimated water fluxes based on chloride data were very low (0.03 to 0.05 mm/yr below 10-m depth).

The above discussion is based on natural conditions. Of course, the facility itself must be designed so as not to alter this naturally low recharge. Additional discussion and recommendations on this point can be found in Chapter 7 dealing with the Wilshire group's Issue #5.

Limitations of Field Data at the Ward Valley Site

The committee notes that monitoring hydraulic parameters in dry soils like those at the Ward Valley site is very difficult and may be one of the causes of the limitations in collecting field data. Limitations of field data during site characterization are grouped into three classes: (1) restrictions imposed by the extremely low water fluxes, which can cause difficulties, for example, in resolving rate and direction of water movement and in collecting adequate water samples for tritium analysis; (2) limitations of the monitoring equipment in arid unsaturated zones, because of the lack of methods, procedures, and reliable instruments to measure precisely the hydraulic and hydrochemical parameters used to estimate water flux in dry desert soils, and because some of the instruments used for the site studies have a high failure rate; and (3) limitation in the quantity and quality of the data whereby the number and distribution of observations and quantity of data collected were restricted. Specifically these include errors in installation and operation of the unsaturated zone monitoring equipment, inconsistencies and errors in the methodologies, analysis, and presentation of data in the license application, and/or project decisions on where, how often, and how deep to test. Detailed discussion of the limitations and inconsistencies of some of the data can be found in

Chapter 3, especially in the sections on the nature, direction and magnitude of water flux, environmental tracers, and evaluation of recharge at Ward Valley.

Recommendations for Issue 1

General Recommendations

- In the opinion of the committee, thick unsaturated alluvial sediments in arid environments such as that found at the Ward Valley site generally represent a favorable hydrologic environment for the isolation of low-level radioactive waste, because of the very small amount of water and very slow rate of water movement throughout most of these unsaturated zones. However, the committee attributes some of the incomplete and/or unreliable data sets that it reviewed to the fact that hydrologic processes in arid regions are characterized by extreme events which do not follow a one-year calendar. For this reason, regulatory and/or budgetary guidelines that permit one-year characterization periods, or other short-duration time frames not suitable for arid-soil characterization, can easily lead to incomplete or ambiguous results. In the committee's opinion, characterization activities should receive priority over arbitrary regulatory timetables, or short time-frame budgetary constraints, particularly in arid regions.

- To guard against deficiencies in characterization and monitoring efforts, and as more emphasis is placed on arid regions for waste disposal, the committee recommends that an independent scientific peer review committee be established to provide oversight early in the permitting process, to assess and suggest improvements in the site characterization plans and monitoring investigations, and to guide the interpretation of the long-term monitoring data. In this way, conflicts in, and other concerns with, characterization data and observations from the unsaturated zone can be resolved as they arise. This recommendation is also discussed later with reference to monitoring.

Specific Recommendations

- As water content and water potential monitoring, tritium analyses, and ground-water levels are proposed for operational and post-closure monitoring, the committee recommends several actions to establish base levels for monitoring, including additional testing for tritium, sampling for chlorine-36 (^{36}Cl) to help resolve the reported tritium found in the unsaturated zone, drilling and sampling of the unsaturated zone from well below the current characterization depth of 30 meters, and further investigation of the apparent vertical hydraulic gradient found between monitoring wells WV-MW-01 and WV-MW-02. Details of these and other recommendations can be found at the end of Chapter 3 and in Chapter 6 on monitoring during operations and post-closure.

****

ISSUE 2 (Issue 1 of the Wilshire group): The committee concludes that shallow subsurface (lateral) flow, as proposed by the Wilshire group, is not a significant issue at the Ward Valley site, because under low-water fluxes (1) the soil carbonate, thought to be a low-permeability horizon causing local ponding, or perching of surface water, is sufficiently permeable to allow water to move predominantly downward, and (2) calculations show that, with a two-percent slope of the layering and soil horizons, lateral flow into the trenches would be insignificant even under a worst-case scenario.

Discussion of Issue 2

Basis for Committee Conclusions

Lateral flow under natural conditions in arid soils depends on several factors. Among them are (1) lateral continuity of a perching (or low-permeability) horizon, (2) the relative permeabilities of the soil horizons, and (3) the slope of the less permeable layer.

Although the limited information available for Ward Valley suggests that some of the shallow subsurface carbonate or "calcrete" horizons are laterally continuous, and less permeable than the surrounding soil, studies from other arid regions and incomplete experimental data from Ward Valley indicate that the permeability of the calcrete is high enough to allow downward movement of water under conditions of low-water content and potentials, and extremely low-water fluxes.

Moreover, at the Ward Valley site, both ancient buried surfaces and the modern surface of the alluvial fan have a slope of only about 2 percent, which Darcy flow calculations indicate is too low to allow significant lateral flow in the unsaturated zone. Under low water-content conditions, the very small downslope gravity component of subsurface flow is negligible compared to the downward diffusion component.

Recommendation for Issue 2

• The committee strongly recommends that conditions that could cause local lateral flow, such as ponding and enhanced percolation through runoff-control structures, be avoided in and immediately surrounding the trenches.

****

ISSUE 3: While there are conceivable, but unlikely, flowpaths for some ground water within Ward Valley to reach the Colorado River, the committee concludes from conservative bounding calculations that, even if all 10 curies (Ci) of plutonium-239 expected in the facility were to reach the river, the potential impacts on the river water

quality would be insignificant relative to present natural levels of radionuclides in the river and to accepted regulatory health standards.

Discussion of Issue 3

Basis for Conclusions

- **Based on limited hydrologic and geologic data and the topographic conditions, in the committee's view, the major part of the ground-water flow beneath the proposed site in Ward Valley appears to discharge at Danby Dry Lake.** It cannot be ruled out, however, that some portion of the ground water passing beneath the proposed site may leave the Ward Valley basin.
- **Although, in the committee's evaluation of the pathways, four of the five postulated routes appeared to be possible, the committee judged that it would not be possible under any reasonable expectation for site characterization to either confirm, or eliminate with absolute certainty, any of the regional bedrock pathways postulated by the Wilshire group.**

Assuming that a ground-water pathway is possible, the committee assessed the potential impact of some concentration of long-lived radionuclides reaching and entering the Colorado River. For the bounding calculation, the committee used a total inventory of about 10 Ci of plutonium-239 (^{239}Pu) emplaced in the disposal site over the 30-year life of the facility, equivalent to 0.33 Ci per year. This order of magnitude was agreed upon by the Nuclear Regulatory Commission and the Congressional Research Service in separate analyses and is less than that proposed by opponents of the site by two orders of magnitude. Radionuclide composition of the waste is certified by the waste generators and will be monitored by the California Department of Health Services and the site operator. DHS has enforcement authority over radionuclide quantity restrictions.

With some overly conservative, non-credible assumptions, and an assumed plutonium release of 0.33 Ci per year, (which would be a release of all the estimated plutonium to be received over the life of the site) *all of which is assumed to reach the river through the ground water,* the calculations showed that the total annual concentration of alpha-emitting radionuclides from Ward Valley ground water that could be added to the Colorado River would be equivalent to a concentration of 0.07 picocuries (pCi) per liter (l). The reported concentration of alpha-emitting radionuclides being transported annually by the river is 44 Ci, equivalent to an average concentration of 4.4 pCi/l, based on recent river flow rates, which are much higher than the future flow rate assumed for the above plutonium calculation. **The committee concludes that the addition of 0.33 Ci per year would be insignificant compared to the natural alpha load of 44 Ci per year carried by the Colorado River. The addition of 0.07 pCi/l from waste plutonium to the existing load in the river of 4.4 pCi/l would result in a total load of 4.47 pCi/l, well below the health-based regulatory standard of 15 pCi/l for gross alpha-emitting radionuclides.** Detailed discussion of the pathways, and of the calculations and assumptions used, can be found in Chapter 5 of this report.

****

ISSUE 4: With respect to the performance monitoring of the unsaturated zone and compliance monitoring of the ground water, the committee concludes that the Wilshire group's concerns for the absence of such plans are not borne out, as the administrative record provides definite plans for post-closure monitoring downgradient in the unsaturated zone beneath the trenches and at the water table at the site boundary. However, although remediation plans are described for ground water contamination, none are described in the revised plan for the unsaturated zone.

Discussion of Issue 4

Compliance and Performance Monitoring

Two basic types of monitoring are proposed for Ward Valley: (1) regulatory compliance monitoring of the ground water to assure that contaminant releases do not exceed regulatory levels at the disposal system boundaries and (2) performance monitoring of the unsaturated zone to provide an early warning of releases that may exceed regulatory levels. The compliance boundaries are the air, vegetation, and water table at the edge of the buffer zone. Monitoring in the unsaturated zone, which constitutes performance monitoring because the unsaturated zone is not a regulatory compliance boundary, is critical because this is the primary natural barrier to radionuclide migration. The data provided will be essential in evaluating the performance of the Ward Valley facility and will be compared with the results of performance assessment models of the site. **Monitoring beneath the trenches for changes in water content and presence of radionuclides in the gas and liquid phases constitutes downgradient monitoring in terms of the unsaturated zone, thus responding to the Wilshire group's concern.**

Integrated Approach

Although traditionally, site characterization, monitoring, and performance assessment have each been conducted independently, various agencies involved in waste disposal, such as the Nuclear Regulatory Commission (US NRC) and the Department of Energy (DOE), increasingly recognize that these three activities should be integrated. More emphasis should be placed on continued site characterization in the proposed operational monitoring programs because reliance is ultimately placed on the natural system as the primary barrier to contain the waste. **In the opinion of the committee, monitoring and performance assessment should** be integrated with continued site characterization during operation of the Ward Valley site.

Conclusions on Saturated Zone Monitoring

- The committee considers that the proposed spacing of monitoring points along the perimeter of the radiological control area may not be adequate and recommends the installation of additional monitoring wells, including wells to examine changes in fluid potential and water quality with depth. In the opinion of the committee, the addition of these wells will provide sufficient monitoring points for ground water, and that no downgradient, offsite wells are required.

Recommendations for Issue 4

Unsaturated Zone Monitoring

- Although fundamental site characterization data are collected prior to the license application, it is the committee's opinion that site characterization should be continued through the operational phase.
- As federal regulations provide investigation and action levels for compliance monitoring, but not for performance monitoring in the unsaturated zone, the committee strongly recommends developing and documenting investigation and action levels for monitoring in the unsaturated zone. In the unlikely event of contaminant movement through the unsaturated zone, it is not prudent to wait for contaminants to reach the compliance boundary before investigating the contaminant movement and developing an action plan.
- Unsaturated-zone hydrology is a relatively young science, and new technologies are continually being developed. In view of the complexities in monitoring unsaturated zones in arid systems and the poor quality of the monitoring data collected for the license application, the committee recommends that future monitoring be directed and overseen by a peer-review advisory panel. It should include experts especially knowledgeable in the state-of-the-art of unsaturated-zone hydrology, soil physics, arid zone water-balance modeling, performance assessments, and ground-water hydrology in arid regions. These experts would also assist in reviewing the monitoring data and in recommending ways of rectifying any problems that may arise. In addition, the peer-review advisory panel should be involved in evaluating iterative processes of site characterization, monitoring, and performance assessment. Continuing scientific peer review would build credibility and public confidence in the monitoring program.

Saturated Zone (Ground-Water) Monitoring

- **The committee recommends that each of the southern and eastern perimeters of the radiological control area have no fewer than four monitoring wells, inclusive of corner monitoring locations (i.e. a total of eight monitoring wells). In addition, to establish better background databases, the western and northern perimeters should be equipped with no fewer than three monitoring points (i.e. a total of three background wells).**

Additional concerns and recommendations with respect to the monitoring plan can be found in Chapter 6 in the sections on performance monitoring and compliance monitoring.

****

ISSUE 5: The committee concludes that the proposed flood protection barrier (berm), which is designed to surround and shield the waste site from flooding and erosion, appears to be effectively engineered with thick stone (rip rap) and gravel (filter) layers to protect the trenches and cover from a rare, desert surface runoff flood event, such as the probable maximum flood (PMF) which is often associated with return periods ranging from 1,000 to 1 million years. Furthermore, any postulated formation of channels by storm water runoff toward the site and resultant scouring around the upstream corners of the flood protection berm appear to be adequately addressed by the rip-rap design above and below ground surface level.

- **The proposed system of shallow ridges (or flow break-up berms) to be built of natural site material and placed upslope from the waste facility site in a chevron pattern, to provide roughness to reduce the velocity of water coming off the fan toward the site, will likely be eroded and breached over a period of decades but will probably continue to function to provide the desired flow resistance for several additional decades, and will have no impact on the stability of the site.**
- **In the committee's opinion, concerns over possible floodwater ponding along the upstream edge of the flood protection berm and possible water seepage through the berm and into the trench area can be effectively ameliorated through easily-engineered defensive measures.**

Discussion of Issue 5

Flood Protection Berm

Offsite storm water will be prevented from entering the trench area during operation and after closure by a permanent flood-protection berm surrounding the disposal site. The berm, to be built when site construction begins and incorporated into the final site cover at closure, is designed to withstand the probable maximum flood (PMF) and also to divert flow around the north and south sides of the facility during operations and after closure. Embankment armoring, consisting of a layer of stone rip rap (0.9 m thick) and filter base of gravel (46 cm thick), is proposed to stabilize the surface against wind and water erosion. The outer embankment armoring system is to be extended to a depth of 1.5 m into the subsurface to provide some scouring protection from adjacent surface water flow.

Breakup Berms

A series of shallow, flow breakup berms will be placed in a staggered, offset chevron pattern upslope and west of the disposal facility primarily (1) to create sheet flow roughness to decrease sheet flow velocity near the permanent primary flood control berm and thereby reduce scour potential and (2) to divert storm runoff to the north and south of the facility. These breakup berms, however, are meant to be temporary and will be constructed with materials removed from the trenches and maintained during the operations and institutional control periods. Although these berms are temporary, and are likely to erode and breach over a few decades, they will remain in some form to offer resistance to sheet flow for many decades after closure.

Recommendations for Issue 5

• **To eliminate the possibility of ponding along the upstream edge of the flood protection berm and to reduce the possibility of infiltration and leakage into the adjacent trench, the committee recommends an engineered sloped and lined channel for conveying storm water around the west, north, and south sides and corners of the flood protection berm. A lined channel/berm would also reduce non-flood event rainfall infiltration and seepage into the trench area.**

• **The committee recommends developing a long-term monitoring plan for detecting significant differential settlement of the trench-cover area and a response program for mitigating its potential negative effect(s) on surface drainage and floods. This plan also should include a comprehensive, operational and long-term flood and erosion-facility monitoring and response program for identifying, repairing, or mitigating any stability problems which develop.**

Additional detailed discussion and recommendations can be found in Chapter 7.

****

ISSUE 6: The committee has two primary concerns about potential effects of the proposed facility on desert tortoise habitat: (1) limited habitat degradation and fragmentation associated with development of the facility, and (2) the unknown

consequences of the relocation plan. **The committee concludes that the parts of the mitigation plan dealing with potential growth of predator populations and increased tortoise/human interactions are likely to minimize these adverse effects, but the plan to relocate tortoises may be detrimental to the tortoise population in the vicinity of the site.**

Discussion of Issue 6

Plan to Remediate Potential Impacts

The site of the proposed facility is in a section of the Mojave Desert that contains one of the largest and most robust desert tortoise populations and is considered to be a vital area for recovery of the desert tortoise. The proposal for the Ward Valley site comprises several approaches for mitigating adverse effects of site construction and operation on the local desert tortoise population. These were initially based on the Bureau of Land Management (BLM) management plan for desert tortoise habitat, and include compensation for lost habitat, reduction of negative impacts on tortoises during facility construction and operation, and research to improve the understanding of desert tortoise ecology. These were also included as Reasonable and Prudent Measures in the USFWS November 1990 Biological Opinion.

Habitat Loss

First, the proposed facility in Ward Valley will result in the direct loss of 36 hectares (ha), disturbance of additional area by road widening and establishment of monitoring equipment, and fragmentation of desert tortoise habitat within the Desert Wildlife Management Area (DWMA) considered by the Recovery Plan to contain the largest and most robust of the remaining desert tortoise populations. While the number of tortoises affected and the habitat area lost are small compared to the whole Ward Valley area, loss of habitat through fencing and road improvement must still be considered habitat fragmentation, a condition that should be avoided according to the US Fish and Wildlife Service (USFWS) 1990 Biological Opinion.

Compensation for Habitat Loss and Fragmentation

The licensee intends to compensate for lost tortoise habitat with a two-part plan: (1) fencing Interstate Highway I-40 and upgrading freeway underpasses to improve habitat currently supporting a low density of tortoises and to facilitate movement throughout Ward Valley; (2) relocating tortoises displaced during site construction into the protected habitat north of I-40 created through fencing along the highway. Approximately 23-30 tortoises will be moved as part of the relocation effort.

• In the committee's opinion, although the relocation site will have geology and soils similar to the waste facility site, the committee sees several problems associated with the relocation plan, some of which are: (1) previous desert tortoise relocation studies have shown only limited success, (2) according to the guidelines for tortoise translocation in the USFWS Desert Tortoise (Mojave Population) Recovery Plan, displaced tortoises should not be released in DWMAs until relocation is much better understood, and (3) the relocation plan could facilitate the transmission of disease from tortoises in the Fenner DWMA to individuals in the Chemehuevi DWMA.

Recommendations for Issue 6

• The committee recommends that the relocation plan be reevaluated in the light of the Desert Tortoise (Mojave Population) Recovery Plan and the paucity of data on successful tortoise relocations.

• The committee supports the recommendation of the U.S. Fish and Wildlife Service to establish a research program designed to study the effects of tortoise relocation and further recommends that relocation be made only outside DWMAs.

• As a possible alternative for the relocation plan, the committee suggests that consideration be again given to (1) evaluating impacts on the adjacent tortoise population of a plan that would exclude, but not relocate, resident individuals from all locations of facility construction and operation activities, or (2) consultation with the U.S. Fish and Wildlife Service about designating all individuals lost during construction as "incidental take".

• The committee recommends that formal consultation with the U. S. Fish and Wildlife Service on the low-level radioactive waste disposal site at Ward Valley be reinitiated, which is required by the Endangered Species Act if critical habitat is designated which may affect a prior biological opinion.

Detailed discussion of the plans and additional problems identified with the relocation plan, and further recommendations, can be found in Chapter 8 of this report in the sections on the assessment of the plan, conclusions, and recommendations.

****

ISSUE 7: In the opinion of the committee, the guidelines presented as part of the revegetation plan have been developed with an understanding of desert plant ecology, and do not reveal any "misconceptions about revegetation enhancements", as charged by the Wilshire group.[4]

---

[4] M. Mifflin dissented from this conclusion. See Appendix F for his views on this issue.

Discussion of Issue 7

The Revegetation Plan

Although no active revegetation program is presently in place, the revegetation plan proposal calls for establishing a comprehensive revegetation program. This program will have three phases: (1) transplanting cacti and yuccas during construction, (2) revegetation of caps of completed trenches during operations, and (3) restoration of the entire site after closure. Qualified biologists will be invited to participate on an *ad hoc* committee to help develop revegetation procedures and criteria for evaluating success of revegetation efforts.

Consequences of Elevated Trench Cover

Because the trench cover will be elevated above the surrounding terrain, a moisture gradient will result from the upper end of the cover (caps) receiving only incident precipitation while the lower end receives some runon from the upper end. This in turn is likely to produce a vegetation cover gradient with the upper end of the trench caps having sparser plant cover and possibly less robust plants than the lower end.

If properly planned and fully restored according to established guidelines and expert input, the vegetation cover gradient should not cause a problem relative to soil erosion because the upper end of the trench cap will receive only rainfall and thus will not be impacted by surface flow erosion. At the lower end of the cover, and in the troughs between, the increased runoff is expected to be compensated by an increased vegetation cover.

Recommendations for Issue 7

• **The committee recommends that, although moisture and vegetational gradients of the raised trench caps are expected, the revegetation program include from the start plantings of native plants designed to produce densities and cover equivalent to that expected in the high density areas, that is, equivalent to the natural desert plant distribution.**

The committee emphasizes the need for continued monitoring of the revegetated areas as part of the long-term monitoring program.

1

INTRODUCTION

In June, 1993, the Secretary of the Interior, Bruce Babbitt, received a memorandum from three United States Geological Survey (USGS) geologists. The geologists, Howard Wilshire, Keith Howard, and David Miller, expressed concern that, in their professional judgment, the site evaluation studies for the Ward Valley, California, proposed low-level radioactive waste (LLRW) facility in the eastern Mojave Desert were inadequate to determine the suitability of the site to isolate the waste and left several critical issues unresolved (Wilshire et al, 1993a). The site studies were done by a contractor to the California Department of Health Services (DHS), the state agency responsible for the licensing and regulation of the LLRW facility, and were accepted by the DHS. A license was issued to U.S. Ecology in September 1993, that has been set aside by the court pending the resolution of several legal challenges.

The USGS disclaimed any role related to the issues raised by the geologists (who will be referred to throughout this report as the Wilshire group) and, moreover, had not participated officially in the site evaluation. In the view of the USGS, the Wilshire group was therefore speaking as individuals and not in any official capacity as USGS employees (USGS, 1993). Nevertheless, opponents of the Ward Valley site urged that the Wilshire group's views be given consideration because they were qualified earth scientists. The Wilshire group subsequently elaborated in a more detailed report on the seven issues that they had briefly summarized in the June memorandum (Wilshire et al, 1993b).

The DHS and their contractor for development of the site, U.S. Ecology (USE), replied that the Wilshire group's arguments lacked scientific merit and that all of the issues they raised had been adequately addressed during the entire 2-year siting process (Romano, 1993). The controversy led the Department of the Interior (DOI), the department under which the USGS operates, to request that the National Academy of Sciences' National Research Council (NAS/NRC) convene a committee of experts to evaluate the seven issues raised by the USGS geologists (see Appendix A). This report is the product of that committee's review.

Department of Interior Involvement

Although the responsibility for storage, disposal, and management of LLRW has been assigned to the states by federal law, the Department of Interior became involved because the Bureau of Land Management (BLM), an agency of the DOI, owns the land on which the site is located. Title of the land must be transferred or sold to the state of California before the site can be developed as a LLRW disposal facility. The DOI wanted the results of the NAS/NRC review, along with other considerations, before making a decision on the transfer.

FEDERAL LOW-LEVEL RADIOACTIVE WASTE POLICY ACT

The effort of the state of California to site a LLRW disposal facility, as with other states across the country, has its origins in the Low-Level Radioactive Waste Policy Act of 1980. This bill gave to the states the responsibility of disposing and managing their commercial LLRW. The 1985 amendment to that act set milestones and incentives for developing such facilities, with penalties if progress and goals were not achieved.

The need for LLRW legislation arose when the last three LLRW disposal facilities operating for commercial wastes decided in the late 1970's that they would no longer continue to accept radioactive waste from the entire country. Hanford, Washington and Barnwell, South Carolina experienced difficulties with corrosion and leaks of waste packages before the 1980 federal act and subsequent regulations governed such activity. As the number of on-line nuclear power plants was increasing, and the use of radioisotopes in medical research and treatment kept growing, the need for disposal capacity for the wastes resulting from these activities, and the need for regulations to protect the health and safety of the public, became more pressing. Nevada closed the Beatty site to low-level waste at the end of 1992. As of July 1, 1994, the two remaining disposal sites were closed to states outside their regional compacts. At present such states, including California, are maintaining their wastes in temporary storage facilities, usually at the locality where the waste is generated, such as university research centers, hospitals, and nuclear power plants.

Regional Compacts

The Low-Level Radioactive Waste Policy Act allowed each state to decide if it would proceed alone or join a group of states in its region to share a facility in fulfilling their responsibilities for providing disposal capacity for non-government LLRW generated within their borders. The regional groups of site-sharing states are called compacts and, under acceptable conditions, are approved by the United States Nuclear Regulatory Commission (US NRC)[1] upon application. There are ten such compacts, two of which have some member states not within their geographic region, and several unaffiliated states that chose independent paths. California, a major generator of radioactive wastes by virtue of its nuclear power plants, of which two are now in operation, and its many university research and medical facilities, formed the Southwest Compact with Arizona, and North and South Dakota. California was designated the host state, responsible for building the first LLRW facility for the compact.

---

[1] Note that the National Research Council and the Nuclear Regulatory Commission have the same initial letters. Throughout this report they will be distinguished by referring to the Council as NAS/NRC and to the Commission as US NRC.

Regulations Governing Low-Level Radioactive Waste

The law also designated the US NRC to provide regulations and guidelines for site selection and safe disposal of civilian radioactive wastes. Among the guidelines in the US NRC's regulations, found in Title 10, Part 61, of the Code of Federal Regulations (10 CFR Part 61), is a definition of LLRW (see Box 1.1) and a description of the types of wastes that are allowable for a LLRW site and classification of radioactive wastes based on the concentrations, half lives of the isotopes, and the types and intensities of activity (See Appendix B). Low-level radioactive waste is broadly defined as any radioactive waste that is not spent nuclear fuel, transuranic waste, or uranium mill tailings.

Classification of Wastes

A classification of radioactive waste was developed by the US NRC. Class A, B, and C wastes, and Greater Than Class C refer to the relative hazard of the radionuclides in the waste. The particular class into which a waste falls is determined by the concentration in the waste of (1) specific short-lived radionuclides (half-lives of weeks to 100 yrs) and/or (2) relatively long-lived radionuclides (half lives of about 500 years or longer) that are below the activities required for classification as high level waste. Class A, B, and C wastes can be disposed of in shallow-land burial trenches or other near-surface facilities. Greater than class C waste, however, must be disposed of in a high-level radioactive waste facility or some facility licensed by the US NRC (US NRC, 10 CFR Part 61) (see Appendix B for details of waste classification).

Composition of Low-level Radioactive Wastes

Low-level radioactive waste may be anything from test tubes, hypodermic needles and animal carcasses to contaminated rags, rubber gloves, tools, decontamination resins and solutions from nuclear power plants, and parts of nuclear power plants other than the core, fuel rods, or other highly active, long-lived radionuclide-contaminated parts. However, the wastes cannot be accepted for disposal as a fluid. In contrast to the way that some low-level radioactive waste has been disposed of in the past, only containers originally packed with dry waste are permitted in a LLRW disposal facility. Dry is defined in 10 CFR Part 61 as containing less than one percent free-standing, non-corrosive liquid. This interdiction results from the knowledge among earth scientists and regulators that ground-water pathways are the most likely way in which radionuclides can reach beyond the disposal site boundaries, and liquid wastes are more likely to find their way to the ground water if the containers fail. Although gaseous releases are also possible, dispersion in air is usually rapid enough that it is considered unlikely to reach the public in harmful concentrations. Potential leaks from containers with fluids are therefore avoided by the requirement for dry material only. Thus the kind of leaks that may have occurred in early disposal sites, whether plumes of contamination within the soil, or contaminated ground water, are considered less likely by federal and state regulators if current federal and state regulations are followed and the sites are managed responsibly.

Siting Considerations

Given the concern about ground water, it has long been recognized that the safest places to store hazardous wastes, especially radioactive waste, would be in the unsaturated zone in desert climates (Winograd, 1974; National Research Council, 1976). In these environments, rainfall is minimal, surface and soil water evaporate rapidly, and plants transpire water vapor to the atmosphere to remove most water within the upper part of the soil or surface layer of earth material. Evaporation and transpiration are referred to jointly as evapotranspiration. Such a condition, it is argued by proponents of areas with desert climates for disposal of hazardous wastes, would prevent accumulation of water in a subsurface trench or other near-surface facility above the water table, and thus minimize the possibility of water and contaminants passing through the unsaturated zone to the water table. However, there are circumstances for which assumptions of dryness in the unsaturated zone in arid climates may be challenged.

Multiple Barriers

Moreover, as the regulations require reasonable assurance that the public will be protected against exposure to releases of radioactivity beyond the established regulatory limits, the siting guidelines encourage multiple barriers to isolate the waste. The natural barrier is regarded as the major line of defense. In the case of Ward Valley, it would be the 180-213 m thick unsaturated zone, that part of the surface material, including the soil, that lies above the water table. The water table, as defined here, is the uppermost surface of the saturated-zone ground water. Recent advances in the understanding of unsaturated-zone processes, and newly developing techniques in analysis, especially in desert climates, have led to considerable confidence in the ability of such a thick unsaturated zone in a desert environment to isolate the radionuclides in a radioactive waste facility (Bedinger et al., 1989; Reith and Thomson, 1992).

In addition, the design of the facility is required to provide redundant protection, and to enhance the ability of the natural barrier to isolate the waste. The cover of the facility, for example, must be designed to minimize infiltration of water which can be accomplished by diverting flow from the waste and by revegetation of native plants on the cover. Any device or construct designed to protect the waste from contact with water is referred to as an engineered barrier. The type of container or waste package is an example of an engineered barrier. Berms or rip rap (barriers of blocks of rock) built up to prevent the effects of energetic surface runoff after a rain, such as during a flash flood, is another example of engineering to protect the facility from erosion that could weaken its first line of defense. Engineered barriers are intended to enhance the natural barrier.

Ground-Water Travel Time as a Barrier

Although it may appear to be a contradiction to what was stated previously concerning the ground water as a pathway to public water supplies, the ground water can be thought of as still another possible barrier to the transport of radionuclides in sufficient concentrations to pose a risk to the public. That is because under some circumstances ground water can move so slowly through earth materials that it may take hundreds or even thousands of years to travel a few kilometers. Moreover, precipitation or adsorption of contaminants may take place along the way. The slow rate of transport would allow decay of shorter-lived radionuclides to extremely low concentration levels below the natural levels found in ground water, referred to as background levels. Depending upon its rate of movement toward the disposal site boundaries and beyond, the ground water could have little or no effect on the concentrations of longer-lived radionuclides. However, the concentrations of a large number of radionuclides, especially shorter-lived species, would be much reduced in their ground-water path beyond the site boundaries because the slow travel time would allow time for radionuclides to decay to stable elements, or to adsorb to minerals in the aquifer through which the ground water flows and thus remove some contaminants.

THE WARD VALLEY CONTROVERSY

The preceding general remarks provide the context within which the debate arose concerning the Ward Valley site in California. The two-year siting and licensing process, which included public involvement from the early stages, was coming to a close after several court challenges by opponents to halt the project. As the decision to transfer the land to the state of California was being considered, DOI received the memorandum from the Wilshire group.

Issues Raised by the Wilshire Group

The memorandum from the Wilshire group described briefly seven concerns that arose from their reading of the draft Environmental Impact Report/Statement (EIR/S). The seven issues, as stated by the authors in the memorandum (Wilshire et al., 1993a), are:

1. Potential infiltration of the repository trenches by shallow subsurface water flow.[2]
2. Transfer of contaminants through the unsaturated zone and potential for contamination of ground water.
3. Potential for hydrologic connection between the site and the Colorado River.
4. No plans are revealed for monitoring ground water or the unsaturated zone downgradient from the site.
5. Engineered flood control devices like those proposed have failed in past decades at numerous locations across the Mojave Desert.
6. Alluvium and colluvium derived from Cretaceous granite appears to make a very high quality tortoise habitat. Sacrifice of such habitat cannot be physically compensated.
7. Misconceptions about revegetation enhancement may interfere with successful reestablishment of the native community.

The Wilshire group provided a brief paragraph of explanation for each of the seven issues (Wilshire et al., 1993a).

The DHS/U.S. Ecology Response

U.S. Ecology (USE), the contractor to DHS, replied to Secretary Babbitt in a letter dated June 25, 1993. In it, USE pointed out that the Wilshire group cited the draft EIR/S to identify the seven issues, a version that had been superseded by the final EIR/S after a comment period in which the public and others provided their reviews of the draft and offered suggestions for improvements. The final EIR/S therefore had incorporated many suggestions from the public comment period. During that time, there was ample opportunity for comment by interested agencies and parties, but the Wilshire group had not participated. The letter asserted that had the Wilshire group become familiar with the licensing record, they would have learned that all the issues belatedly enumerated by them had already been addressed.

---

[2] This refers to subsurface *lateral* flow as confirmed by the Wilshire group.

The Second Report

The Wilshire group replied with a second more detailed report dated December 8, 1993. They elaborated on the first five issues, modifying the issues after reading more up-to-date sections of the administrative record than the draft EIR/S, including the final EIR/S. They did not comment further on the ecological issues of the desert tortoise habitat and revegetation. In the second report they added some information and concerns about the geochemistry, including tritium occurrences in the unsaturated zone and dating of the ground water by the carbon-14 (^{14}C) method. The Wilshire group also postulated five specific pathways by which ground water might flow from Ward Valley to the Colorado River (Wilshire et al., 1993b).

The NAS/NRC Committee's Review

In response to a letter of request from Secretary Babbitt, dated March 14, 1994, the NAS/NRC convened a volunteer committee of experts with training, research, and field investigation experience in the specialized disciplines necessary to evaluate the seven scientific issues. The committee included two unsaturated-zone hydrologists, one soil physicist, four ground-water hydrologists, three geologists, one geophysicist, one flood control/civil engineer, three geochemists, and two ecologists.

Limits of NAS/NRC Review

The NAS/NRC, at the request of the DOI, agreed to review only the specific issues identified by the Wilshire group. Based on the agreement reached with DOI, the Committee to Review Specific Scientific and Technical Safety Issues Related to the Ward Valley, California, Low-Level Radioactive Waste Site, the official name of the NAS/NRC committee, reviewed the literature and other relevant material related to the seven issues to (1) assess the adequacy of the site studies relative to the enumerated issues and the validity of the conclusions concerning site performance that are the subject of the debate, and (2) determine if the enumerated concerns have merit and, if so, to assess the impacts on site performance.

Source of Documents and Data

To accomplish this, the committee reviewed many parts of the administrative record (AR) and other reports and documents relevant to the issues. The AR includes the license application (LA), the DHS interrogatories and USE responses relative to that application, reports related to siting, and data from the monitoring of the site. The Wilshire group also submitted other information to support their position on the issues.

In addition, because of the great public interest in the Ward Valley issue, the committee invited all interested parties to submit written information that they thought would help the committee in its review and deliberations. The committee made it clear that its review was restricted to the seven issues raised by the geologists and that, despite urging from many individuals and organizations, the committee could not consider wider issues or issues other than those scientific and technical issues that NAS/NRC had contracted to review.

Time Constraints

The request for a limited and short-term review of the issues put a great time constraint on the committee which very quickly became inundated with information: 5000 pages selected from the AR, dozens of documents of characterization and monitoring data and reports from USE, new documents generated by the Wilshire group or others who supported their efforts, several reports from proponent and opponent organizations, and data from regulatory bodies. The effort to get through all of the information in as short a time as possible with no important piece of information omitted in the review was to be followed by an intense deliberation and writing period, in order to meet the requested time frame for the issuance of the report. However, concerned with the possibility that they may be sacrificing thoroughness and careful deliberation for an arbitrary or unrealistic deadline, the committee required more time to write its report than was originally allotted. Three executive sessions were held for deliberation and writing after the open information-gathering meetings.

The Open Meetings

The committee met twice for three days in open session in the city of Needles, California, the closest population center to the Ward Valley site. The Wilshire group and the license applicant and licensor, USE and DHS, were given the time they requested to present arguments and evidence in support of their positions. The Wilshire group presented their views and analyses of the site studies, with supporting information from the surrounding area and local experts. USE and DHS presented data, plans, and explanations to support their conclusions that the issues raised had already been adequately addressed. In addition, independent experts with relevant experience and information on the issues were invited, some by the committee, and some by one side or the other. Those who were not direct parties to the dispute or their invitees, both individuals and organizations, were invited to speak at an open microphone at the end of each day's technical session.

The Beatty, Nevada, Site as Analog

One team of experts invited by the committee were USGS hydrogeologists David Prudic and David Nichols, who have spent several years investigating the Beatty, Nevada,

LLRW disposal site that closed in 1992. It had been in operation for at least 30 years, mostly through the period prior to the promulgation of federal regulations for LLRW siting, operations, and waste form. The studies by Prudic and Nichols took place outside the site boundary to evaluate the behavior of the unsaturated zone in an arid climate, and to test monitoring methods. Prudic and Nichols regard the Beatty site as, in some respects, an analog for the Ward Valley site, as it is located in a similar type of terrain, with similar climate and hydrologic characteristics. There are, however, some uncertainties because of unexplained anomalies in the 30-year monitoring records of well data. Although the natural settings are similar, there are major differences between the Beatty site and the proposed Ward Valley site regarding the types of waste disposed and disposal methods. For example, the Beatty facility includes both a *toxic chemical waste site* and a low-level radioactive waste site adjacent to each other. There are also serious questions or uncertainties regarding the types, compositions, and physical forms of wastes that were accepted at the Beatty site, such as how much toxic chemical and/or radioactive waste were disposed of in *liquid* form. Another characteristic of the Beatty site is that huge disposal trenches were excavated and kept open for many years until filled, allowing accumulation and infiltration of rainfall. The committee therefore decided that the Beatty site may be useful in understanding some natural processes, but it is limited in evaluating the behavior of the Ward Valley site because of the historical uncertainties.

The Plutonium-239 Issue

In the committee's effort to understand the timespan of concern for the facility, it requested information on the likely inventory of radionuclides for the facility. The committee was at that time unaware that the potential amount of plutonium-239 (^{239}Pu) was a controversial issue, not between the Wilshire group and the DHS, but between another organization opposed to the Ward Valley site and DHS (Committee to Bridge the Gap, 1994).

The controversy came about because the U.S. Ecology performance assessment calculations assumed 3500 curies of ^{239}Pu, based on guidelines from the US NRC. Subsequently, experience at nuclear power plants suggested that the amount of ^{239}Pu that would result from the process that produces the ^{239}Pu was two orders of magnitude less than the original assumption. As a result DHS has since maintained that at most "a fraction of a curie to two curies of ^{239}Pu from decontamination waste, and a fraction of a curie to several curies from other sources" would be emplaced over the life of facility operation (Brandt, 1994). Opponents insisted that the larger number is correct.

As the committee needed a reliable estimate in order to assess the potential impact of long-lived radionuclides reaching the Colorado River, it requested an evaluation of the positions of the DHS and Committee to Bridge the Gap, from an independent source, the US NRC. In addition, the committee learned that the Congressional Research Service (CRS) had done a similar study (Holt, 1994). The results of the two organizations' studies converged independently (Holt, 1994; US NRC, 1994a and 1994b), both agreeing that the appropriate estimate is two orders of magnitude lower than that used for the performance assessment.

(The CRS is presently doing a second study because of newer available data). These studies were especially helpful in providing the guidance the committee needed to select an order of magnitude to do some bounding calculations. The committee, therefore, based the order of magnitude for ^{239}Pu expected for such a facility on these two studies in doing the bounding calculations found in Chapter 5 of this report.

Organization of the Report

After this introduction to the study, the report begins with an overview of the geology, geophysics, geomorphology, hydrology, and ecology of the site and surrounding region in Chapter 2. The following chapters cover the seven issues of the USGS geologists. The format of each chapter follows a pattern. First is a statement of the issue, followed by a summary of the Wilshire group's position, the USE and DHS response or position, and the committee's discussion, analysis, observations, conclusions, recommendations and references cited. Chapter 2, then is the site overview; Chapter 3 addresses the potential for water and contaminants to be transferred through the unsaturated zone to the ground water; Chapter 4 deals with lateral subsurface flow in the unsaturated zone; Chapter 5 analyzes the five postulated pathways that ground water could take to flow from the site to the Colorado River; Chapter 6 discusses the monitoring plans for the unsaturated zone and the ground water; Chapter 7 evaluates the flood control devices and other engineered structures to prevent flooding and erosion at the site; Chapter 8 considers the Desert Tortoise and its habitat; and Chapter 9 evaluates the revegetation plan to reestablish the native vegetation at the site after construction.

Note that the sequence of treatment of issues 1 and 2 of the original Wilshire group memorandum has been reversed in this report, with Chapter 3 addressing the second of the enumerated issues, and Chapter 4 addressing the first. Thus, the Wilshire group's issue 2, the potential downward transport of contaminants through the unsaturated zone, is the committee's issue 1 in Chapter 3, and the Wilshire group's issue 1, the potential infiltration of the trenches by shallow subsurface lateral flow, appears as the committee's issue 2 in Chapter 4.

The reason for the reversal was the need to introduce many of the concepts and characteristics of a thick arid-region unsaturated zone in addressing potential downward flow of water through the unsaturated zone to the water table. The committee, therefore, agreed that the discussion of this issue was best presented first. The discussion of shallow subsurface flow of water that follows in Chapter 4 requires far less information about the unsaturated zone and draws upon information already provided in Chapter 3 as background.

COMMITTEE DECLARATION

The committee was not asked to, and did not, evaluate the suitability of the Ward Valley site for a low-level radioactive waste facility. The charge of the committee

(see Appendix A) was to evaluate the validity of the issues raised by the Wilshire group and their implications for the behavior of the unsaturated zone and for the other features of concern. The committee depended upon the information provided by the opposing groups, DHS/USE and the Wilshire group, any relevant information provided by outside interests, the scientific literature, and their own professional judgment. None of the conclusions should be read as either an endorsement of, nor condemnation of, the Ward Valley site. Approval of the Ward Valley site is the responsibility of those government agencies and officials entrusted to make such decisions.

Recommendations offered herein by the committee, using its professional judgment, should not be construed as an endorsement of the site. The recommendations are made in the spirit of enhancing the scientific program and of decreasing the uncertainties in the earth science information, *if the site is developed for a low-level radioactive waste disposal facility.*

REFERENCES

Bedinger, M. S., Sargent, K. A., Langer, W. H. 1989. Studies of the geology and hydrology of the Basin and Range province, southwestern United States, for isolation of high-level radioactive waste - characterization of the Sonoran region, California. U.S. Geological Survey Professional Paper 1370-E, 30 p.

Brandt, E. 1994. Letter to I. Alterman, dated September 22.

Committee to Bridge the Gap. 1994. Informal presentation to the National Academy of Sciences on the proposed Ward Valley radioactive waste disposal facility. September 1.

Holt, M. 1994. Plutonium Disposal Estimates for the Southwestern Low-Level Radioactive Waste Disposal Compact. Memorandum dated June 3. Congressional Research Service. Library of Congress. Washington, D.C.

National Research Council. 1976. The Shallow Land Burial of Low-Level Radioactive Contaminated Wastes. Washington, D.C.: National Academy of Sciences. 150 pp.

Reith, C.C. and B.M. Thomson. 1992. (eds.) The Disposal of Hazardous Materials in Arid Ecosystems. University of New Mexico Press. Albuquerque.

Romano, S.A. 1993. Letter to Secretary Bruce Babbitt, dated June 25.

U.S. Geological Survey. 1993. Letter from R. M. Hirsch, Acting Director, to Senator Barbara Boxer, October 8.

U.S. Nuclear Regulatory Commission. 1994a. Letter from M. R. Knapp to Ina Alterman, National Research Council. Dated November 23.

_____ 1994b. Letter from M.R. Knapp to Ina Alterman, National Research Council. Dated December 14.

Wilshire, H. G., K. A. Howard, and D. M. Miller. 1993a. Memorandum to Secretary Bruce Babbitt, dated June 2.

_____ 1993b. Description of earth science concerns regarding the Ward Valley low-level radioactive waste site plan and evaluation. Released December 8.

Wilshire, H. G., K. A. Howard, and D. M. Miller. 1994. Ward Valley proposed low-level radioactive waste site. A report to the National Academy of Sciences.

Winograd, I. J. 1974. Radioactive waste storage in the arid zone EOS. Transactions of the American Geophysical Union. 55(10):884-894.

2

SETTING OF THE WARD VALLEY SITE

LOCATION

The proposed Ward Valley low-level radioactive waste (LLRW) disposal facility is in the arid eastern Mojave Desert of southeastern California, approximately 30 km[1] west of the city of Needles (Figure 2.1). Ward Valley trends north-south for about 85 km. It is bounded on the west by the Piute, Little Piute, and Old Woman Mountains and on the east by the Sacramento, Stepladder, and Turtle Mountains. The axis of Ward Valley slopes gently to the south, ranging in elevation from more than 650 m near the north end and the flanks of the valley to a low of 186 m at the southern terminus of the valley at Danby Dry Lake. The valley is drained by Homer Wash, which flows south and discharges into Danby Dry Lake (Figure 2.2). Homer Wash and Danby Dry Lake are both generally dry except during and immediately after heavy rainfalls.

The proposed facility is located about 20 km from the northern end of Ward Valley, approximately 1.5 km south of Interstate 40 in the northeastern quarter of Section 34, Township 9 North, Range 19 East, San Bernardino Baseline and Meridian. The site is about 760 m west of Homer Wash (Figure 2.3), on a broad, low-relief alluvial surface that slopes gently (2 percent grade) eastward from the Piute Mountains toward Homer Wash. The site is at an elevation of 650 m, 17 m above the calculated 100-year flood in Homer Wash. The 100-year flood is the level to which flood waters are statistically predicted to rise an average of once in a hundred years. The distance from the proposed site southward to Danby Dry Lake is approximately 65 km.

GENERAL FACILITY DESCRIPTION

The disposal site, as defined by the Nuclear Regulatory Commission in 10 CFR Part 61, includes the LLRW disposal unit area, or "radiological control area", and a surrounding buffer zone. The State of California would control an approximately square area of about 405 hectares (ha) (1000 acres) consisting of Section 34, the south half of the south half of Section 27, the west half of the west half of Section 35, and the southwest quarter of the southwest quarter of Section 26, as shown in Figure 2.4.

The proposed radiological control area, surrounding buffer zone, and a facilities area are all contained within Section 34 and the 405 ha (1000 acres).

---

[1] The metric system is used for measurements throughout the report. Because of common usage, not all measurements were converted from English units. For details of the conversion from English to metric units, please refer to Appendix C.

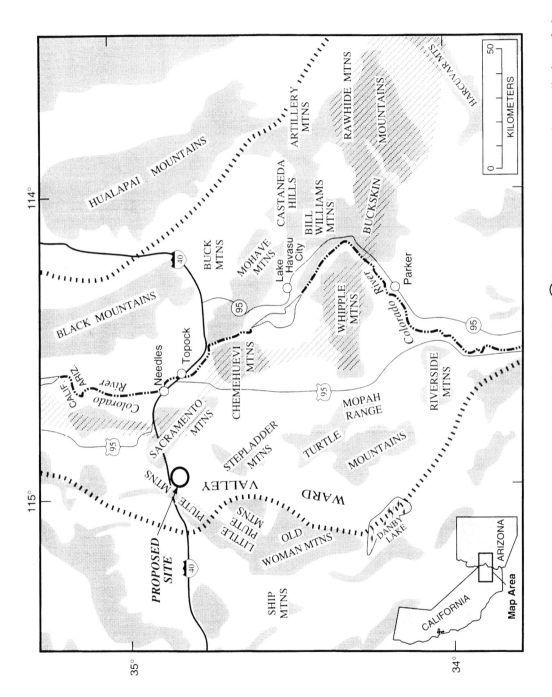

Figure 2.1 Location map of Ward Valley proposed facility site (◯), showing the approximate limits of the Colorado River extensional corridor (thick dashed lines). Shaded areas are mountain ranges. Metamorphic core complexes are marked by diagonal linear patterns. (Modified from Nielson and Beratan, 1995, as cited in Wilshire et al., 1994).

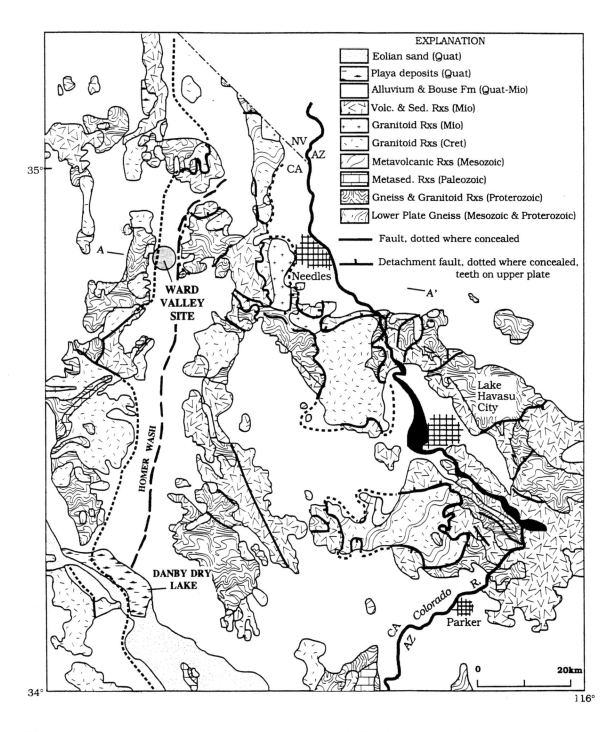

Figure 2.2 Geologic map of the region around Ward Valley. A-A′ designates line of section of Figure 2.6. (Wilshire et al., 1994)

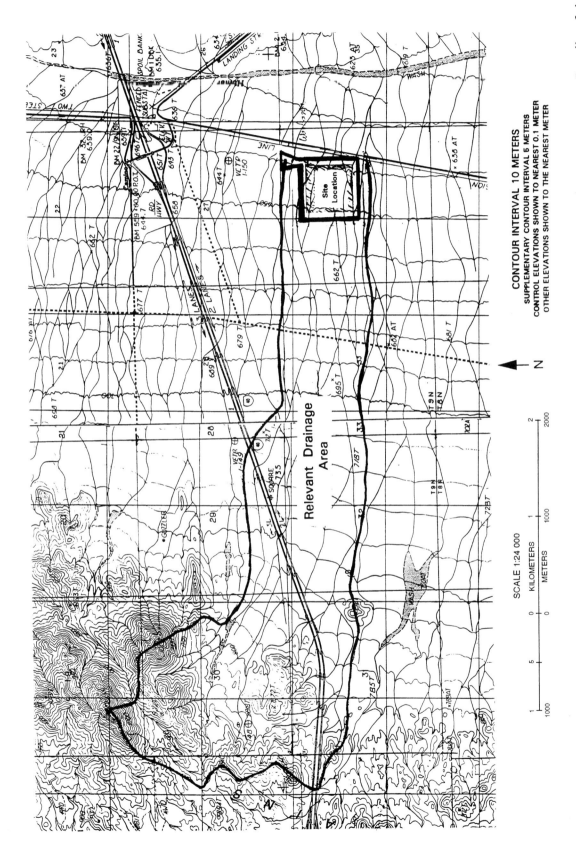

Figure 2.3   Topographic map with location of the site, Homer Wash, the relevant drainage area, and Interstate I-40.   Details of the site location are found in Figure 2.4.   (modified from USGS, 1984)

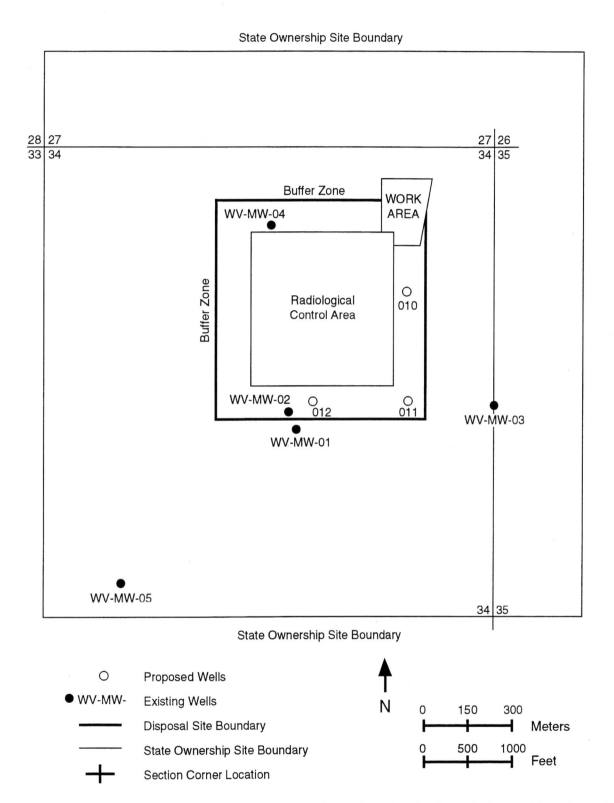

Figure 2.4   Relationship of various proposed Ward Valley site boundaries, and location of proposed and existing monitoring wells (after LA Figure 2420-4).

The proposed disposal facility consists of the following major elements (Figure 2.5):

• A 532 m x 532 m, approximately 28 ha (70-acre), radiological control area surrounded by an electrified security fence, within which near-surface LLRW disposal operations will take place in a series of five progressively developed trenches. Four Class A waste trenches and one Class B/C trench are planned within a 471 m x 471 m trench area. Within the security fence zone, the waste trench area is surrounded by a permanent flood protection berm 1.5 m high on the western upgradient side and 0.9 m high on the eastern downgradient side. The buffer zone will extend 122 m around the entire perimeter of the 532 meter-square fenced control area for carrying out environmental monitoring activities, maneuvering construction equipment, and allowing corrective actions to be implemented. Additional details of the radiological control area and waste trench features are described in Chapter 7.

• An adjoining fenced facilities area of about 3.1 ha (7.6 acres) in the northeast corner of the disposal site containing vehicle parking and staging areas, fuel and water tanks, utilities, a sanitary waste disposal system, a facility operations building, a shop building, test trenches, and a security guard house located at the facility's only entrance gate;

• A series of 0.3 m high drainage berms (called "breakup berms") to the west, upgradient of the disposal area, for increasing the surface roughness and for temporarily diverting shallow surface flow around the control site during construction;

• A meteorological and air-quality monitoring station approximately 90 m southwest of the southwest corner of the facility's security fence;

• Other site elements, including five ground-water monitoring wells and additional environmental monitoring stations (Figures 2.4 and 2.5).

REGIONAL GEOLOGIC SETTING

The geologic history of the eastern Mojave Desert spans more than 1.7 billion years (Howard et al., 1987; Wooden et al., 1988; Wilshire et al., 1994) and is very complex, reflecting multiple episodes of tectonism[2], or large-scale movements, disruptions and heating of the earth's crust, such as faulting, folding, mountain building, metamorphism, and magmatism. The tectonic history can be subdivided into five major phases:

• Early to Middle Proterozoic[3] (from about 1.1 to about 1.8 billion years ago (Ga)) deposition, deformation, metamorphism, and magmatism, resulting in the formation and stabilization of continental crust;

• Latest Proterozoic, Paleozoic to early Mesozoic sedimentation on a stable shelf;

• Middle through late Mesozoic arc magmatism, tectonism, and regional uplift;

---

[2] See Appendix D, Glossary of Terms, for definitions of geologic and technical terms.

[3] See Table 2.1, Geologic Time Scale, for definitions of geologic time periods.

33

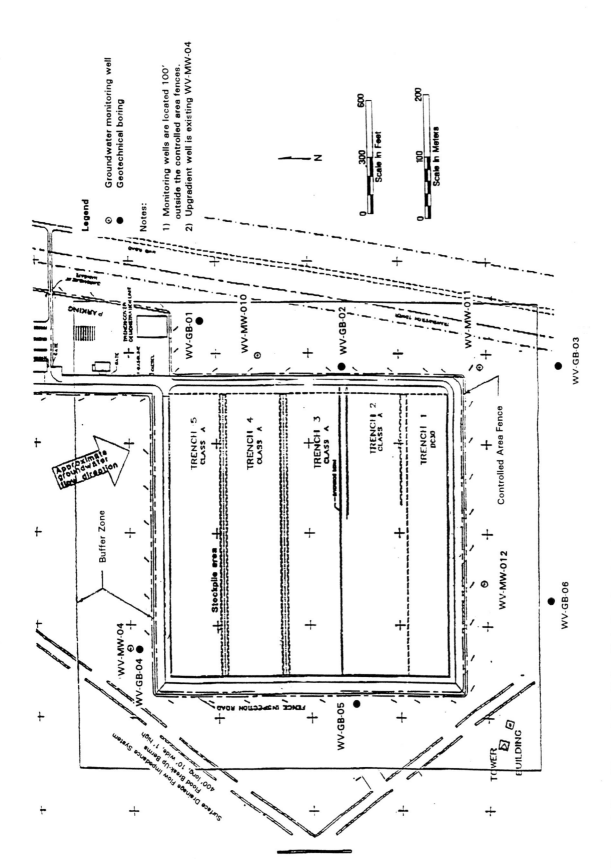

Figure 2.5 The main facilities (LA Section 4400).

Table 2.1   **Geologic Time Scale**

Era	Period	Epoch	Beginning of epoch or period (millions of years ago)
Phanerozoic Cenozoic	Quaternary	Holocene	0.01
		Pleistocene	1.6
	Tertiary	Pliocene	5
		Miocene	24
		Oligocene	37
		Eocene	58
		Paleocene	66
Mesozoic	Cretaceous	Late (upper)	
		Early (lower)	144
	Jurassic	Late (upper)	
		Middle	
		Early (lower)	208
	Triassic	Late (upper)	
		Middle	
		Early (lower)	245
Paleozoic	Permian		286
	Pennsylvanian		320
	Mississippian		360
	Devonian		408
	Silurian		430
	Ordovician		505
	Cambrian		570
Precambrian Proterozoic			2500
Archaen			4550(?)

• Cenozoic (Miocene) crustal extension and broadly synchronous sedimentation and magmatism; and

• Late Cenozoic post-extension erosion of the mountain ranges, filling of the intervening basins with the resulting sediments, and partial integration of drainage systems during a time of relative tectonic stability.

The three pre-Cenozoic phases, i.e. before 66 Ma, produced much of the complex geology found in the mountain ranges (Figure 2.2), including fracture permeability of the basement rocks, but otherwise have relatively minor implications for assessing the proposed Ward Valley site. In contrast, the two Cenozoic phases, i.e. Miocene crustal extension, sedimentation, and magmatism, and late Cenozoic erosion, produced the present-day surface appearance, or physiography, of mostly north- and northwest-trending mountain ranges surrounded by sediment-covered basins, and have much greater relevance to the site.

PRE-MIOCENE GEOLOGIC HISTORY

Proterozoic Era

The oldest rocks in the region are Proterozoic metamorphic and plutonic rocks that were subjected to extremely high temperatures and pressures resulting from great tectonic forces at about 1.7 Ga (Wooden et al., 1988; Wooden and Miller, 1990; Foster et al., 1992). These rocks, exposed in the Piute and Sacramento Mountains, record the formation and subsequent stabilization of the region's continental crust. One of the most extensive Proterozoic rock units in the region is the 1.68-Ga Fenner Gneiss, a coarse-grained biotite granite gneiss that forms dark-colored outcrops throughout the Piute and northwest Sacramento Mountains (Miller et al., 1982; Wooden and Miller, 1990; Bender, 1990; Karlstrom et al., 1993). Other Proterozoic units include Middle Proterozoic syenite (1.4 Ga) and diabase (1.1 Ga) in the Piute Mountains (Howard et al., 1987).

Paleozoic Era

Regional uplift and erosion during Middle to Late Proterozoic time brought the deeply formed metamorphic and plutonic rocks to the surface, where they were eroded to a low-relief surface. These rocks subsequently were overlain by approximately 1 to 1.5 km of Paleozoic and Early Mesozoic (Triassic) sedimentary rocks of similar age and composition as those exposed in the Grand Canyon (Stone et al., 1983). The sedimentary rocks are mostly sandstone, shale, and limestone that were deposited in a shallow sea and accumulated on a tectonically stable part of what was then near the coast of western North American. Small, discontinuous remnants of these Paleozoic and Triassic strata, which were deformed and metamorphosed later, are exposed in the Piute and Little Piute Mountains, west of Ward Valley (Howard et al., 1987; Hoisch et al., 1988; Fletcher and Karlstrom, 1990).

Mesozoic Era

During Late Triassic through Cretaceous time, southern California was the site of intense deformation and granitic magmatism caused by the convergence of tectonic plates along the west coast. Near Ward Valley, this convergence resulted in regional compression which caused thrusting and metamorphism of the Triassic, Paleozoic, and older rocks. Mesozoic granitic plutons invaded the region on the eastern fringes of the Mesozoic batholith belt that includes the Sierra Nevada and Peninsular Ranges batholiths (Miller et al., 1992). Light-colored Mesozoic granitic plutons, such as those exposed in the Piute, Stepladder, and Old Woman Mountains (Howard et al., 1987), were emplaced in the Late Cretaceous and then rapidly cooled and uplifted (Foster et al., 1989, 1990, 1991, 1992).

A lack of Early Tertiary sedimentary or volcanic rocks in the region, in conjunction with evidence from adjacent areas (Bohannon, 1984; Reynolds et al., 1988; Christiansen and Yeats, 1992), suggests that the region remained a topographically positive area, i.e., an upland rather than a basin, from the end of the Mesozoic throughout the first half of the Cenozoic Era.

Cenozoic Era

Miocene Extension and Magmatism

The present-day basins and ranges in the Ward Valley area formed when the earth's crust in the region was pulled apart, or extended, in Miocene time (Howard et al., 1982; Spencer, 1985; Howard and John, 1987). Ward Valley lies along the western side of a 50- to 100-km wide belt of extreme crustal extension that trends northward along the Colorado River (Figure 2.3 and 2.6). Along this belt, Miocene extension stretched the crust to twice its original width in an east-west direction (Howard and John, 1987). Extension began prior to 20 Ma, had largely ended by 13 to 12 Ma (Spencer, 1985; Hillhouse and Wells, 1991; Foster et al., 1990), and was accompanied by volcanism and the emplacement of dikes, sills, and small plutons.

Block Faulting

The extensional event left the region with two very different types of geologic terrain separated by one or more gently dipping regional faults. The first type of terrain consists of north- to northwest-trending fault blocks, in which highly tilted Miocene volcanic and sedimentary strata rest on older rocks. Examples of such fault blocks are present in the Stepladder and Turtle Mountains.

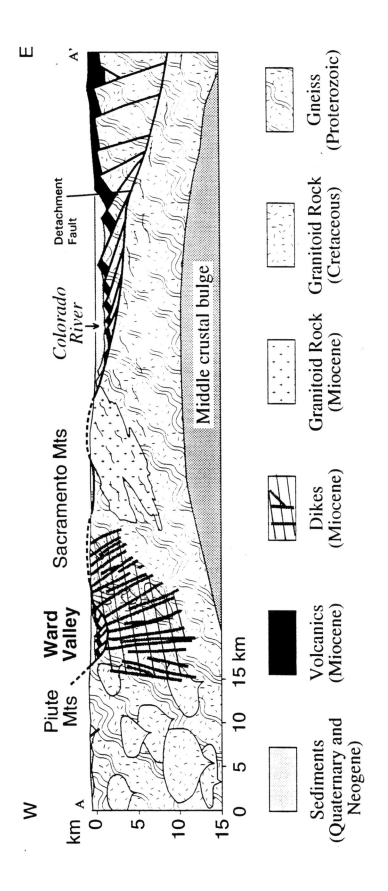

Figure 2.6 Interpretive geologic cross section the Ward Valley site and the Colorado River valley at Needles. Faults shown with heavy lines, dotted where inferred (Wilshire et al., 1994).

The second very different type of terrain consists of deeply formed plutonic and metamorphic rocks that are relatively unfaulted, but instead were heated, sheared, extensively fractured, and contorted in Miocene and earlier times. These deep-level rocks are now exposed at the earth's surface in a belt of mountain ranges, termed metamorphic core complexes. They border the west side of the Colorado River and include, from north to south, the Dead, Sacramento, Chemehuevi, and Whipple Mountains (Figure 2.1).

Detachment Faulting

Separating the two terrains are one or more major normal faults, termed detachment faults. These formed as a thick slab of the upper crust slipped over the underlying crust along a gently sloping (or dipping) fault surface, which in this case sloped toward the east (Figure 2.7). Warping of the detachment faults during crustal extension gave them a dome-like shape over the core complexes (Figure 2.8). The faults are exposed on the flanks of the core-complex ranges, where they juxtapose the tilted fault blocks and associated Miocene volcanic and sedimentary rocks against the underlying deep-level rocks within the ranges (Figure 2.8).

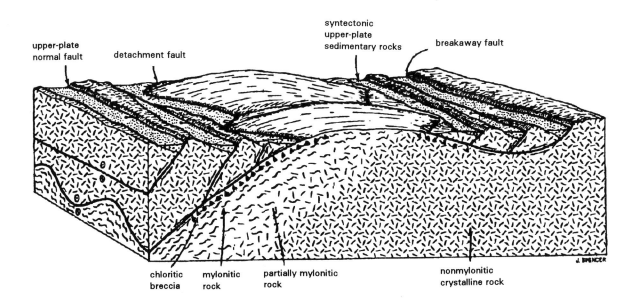

Figure 2.7   Block-diagram of a typical metamorphic core complex and detachment fault, showing the domal form of the complexes and the titled fault blocks above the fault. Greater amount of fracturing typical of upper plate rocks shown by block faulting (Spencer and Reynolds, 1989).

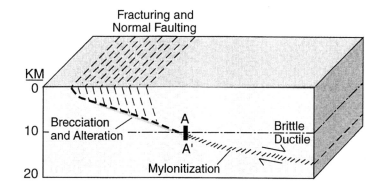

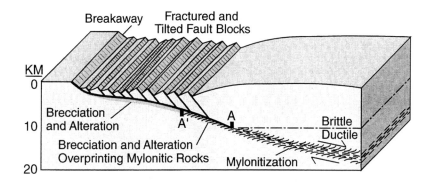

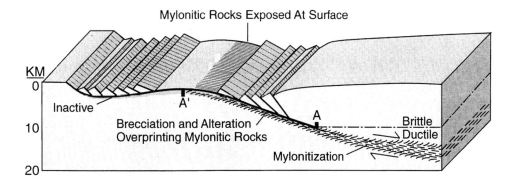

Figure 2.8   Interpretive model for the evolution of detachment faults and metamorphic core complexes. Normal displacement along the low-angle shear zone produces mylonitic (ductile) fabrics at depth and more brittle (fracturing) structures along shallower segments of the fault. As the lower plate becomes less deeply buried with time due to tectonic removal of the overlying rocks, mylonitic rocks are overprinted by successively more brittle structures. Isostatic adjustments lead to the warped, domal geometry of the detachment fault and mylonitic fabrics. In this model the Ward Valley site would be located above the detachment near the word "inactive" on the left side of the bottom figure. The Sacramento Mountains are represented by the mylonitic rocks exposed at the surface (Reynolds et al., 1988).

The detachment faults are interpreted to be present deeper in the crust beneath much of the region (Howard et al., 1982; Spencer, 1985; Howard and John, 1987; Okaya and Frost, 1986), including beneath the proposed LLRW site (Figure 2.9). Most detachment faults are marked by zones of fractured and crushed rock several tens of meters to hundreds of meters thick. During the faulting, hot, chemically rich, ground water passed through and altered the rocks, depositing minerals in the fractured zones as it cooled (Reynolds and Lister, 1987; Roddy et al., 1988; Spencer and Welty, 1989).

A model for the evolution of detachment faults and metamorphic core complexes is shown in Figure 2.8 (Wernicke, 1981; Davis et al., 1986; Howard and John, 1987; Reynolds et al., 1988). In this model, the upper 10 to 15 km of the crust exhibited brittle behavior during extension and broke up into large blocks bounded by normal faults (Figure 2.7). In contrast, rocks in the middle and lower crust, below approximately 15 km, were hot enough, or compressed enough by the overlying crust, to accommodate the extension by ductile, or viscous, flow. These mid-crustal rocks were uplifted to the surface by tens of kilometers of normal displacement along the east-dipping detachment fault. In essence, the deep-level rocks in the California core complexes were pulled out from beneath upper-crustal rocks of western Arizona (Reynolds and Spencer, 1985; Davis et al., 1986; Howard and John, 1987; Davis, 1988; Davis and Lister, 1988).

According to this model, faulting was accompanied by tilting of the blocks, leaving some on end. The Miocene basins formed on the down-dropped parts of the blocks and were filled with volcanic and sedimentary material, including huge landslide rock masses derived from uplifted parts of the blocks. The doming, or upwarping, of detachment fault surfaces is interpreted to have occurred during extension and to be an isostatic response to tectonic unroofing (Spencer, 1985; Howard et al., 1982; McCarthy et al., 1991). Tectonic unroofing refers to the removal of a large overlying rock mass by slippage down a detachment fault. The removal of this dense mass allowed the underlying rock to rebound upward to some equilibrium level, analogous to the way in which a floating block of wood, pushed underwater by a weight, will bounce back to its equilibrium floating position when the weight is removed. The rise of the depressed rock after removal of its "roof" is referred to as isostatic rebound.

The detachment fault may be warped up and over the Sacramento Mountains (Figure 2.9). Alternatively, the detachment beneath Ward Valley could continue dipping gently to the east and go beneath the Sacramento Mountains. In this latter case, the detachment fault exposed in the Sacramento Mountains would be a structurally higher detachment that is not continuous with the fault that underlies the proposed site.

Present-day ranges in the region, therefore, represent (1) single huge tilt blocks or series of tilt blocks above the regional detachment fault (Figure 2.7), (2) uplifted domes of deep-level rocks beneath the warped fault (Figure 2.8), or (3) untilted fault blocks outside the belt of major extension (Davis et al., 1980, 1986; Howard et al., 1982; Howard and John, 1987; Spencer and Reynolds, 1989).

41

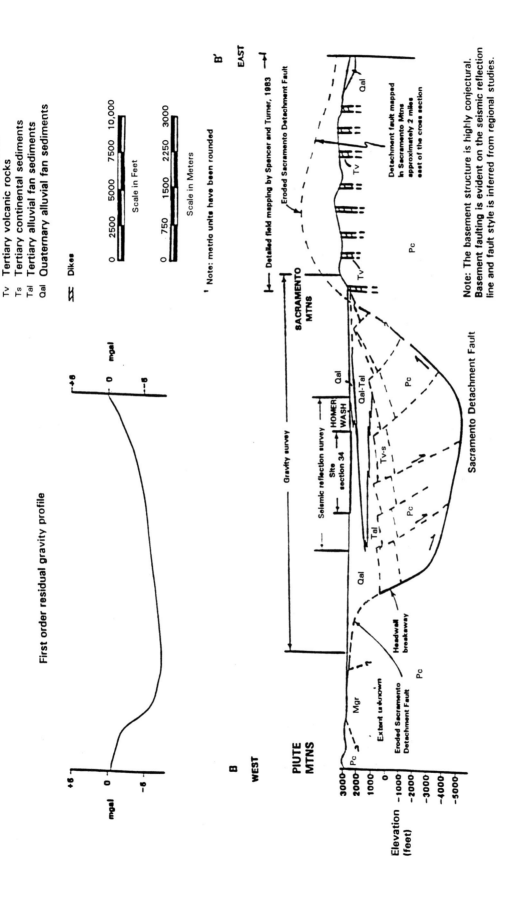

Figure 2.9 Interpretive cross section of the Ward Valley site based on geological, geophysical, and hydrologic data. Section location is on LA Figure 2310.1, approximately the same location as Figure 2.6 (LA Section 2310).

Post-extension Geologic Evolution
and Landscape Development

Late Faults

Major extension and detachment faulting ended between 15 to 12 Ma based on the age of gently dipping volcanic flows that either overlap the detachment fault or unconformably overlie highly tilted Miocene strata in the Sacramento Mountains and elsewhere in the region (Spencer, 1985; Howard and John, 1987; Simpson et al., 1991). A younger, post-detachment reverse fault cuts 12 Ma basalt and conglomerate along the west side of the Dead Mountains (Spencer, 1985). Regionally, these faults do not cut latest Tertiary and Quaternary surficial deposits and are probably middle to late Miocene in age. Although a few minor fault scarps have been identified cutting post-Miocene deposits near Topock, Chemehuevi Valley, Danby Dry Lake, and other locations within 80 km of the proposed facility (Wilshire et al., 1994), at the present time Ward Valley lies within one of the most seismically inactive parts of southern California (Figure 2.10).

Basin Fill

During and subsequent to detachment faulting, erosion of the mountain ranges shed clastic, or fragmented rock, material into the flanking intermountain basins. Late Miocene and younger basin-fill strata dip gently throughout most of the region and unconformably overlie the tilted mid-Miocene and older rocks. Most of the rock fragments are locally derived and can be traced to the adjacent ranges (Metzger and Loeltz, 1973; Dickey et al., 1980; Carr, 1991). The basin fill commonly contains boulders and coarse gravels adjacent to the mountain fronts and becomes progressively finer grained toward the center of the basin (Figure 2.11). The size of clasts, or rock fragments, may also vary as a function of age of the deposit, largely due to variations in climate during the Pleistocene (Dickey et al., 1980; Menges and Pearthree, 1989; Demsey, 1989). Basin-fill and alluvial deposits commonly are relatively unconsolidated near the surface and become more consolidated, or hardened, with depth from compaction and precipitation of mineral material in the pores between the grains.

Some basin fill deposits contain halite (salt) and other minerals that form by evaporation of the water in which the chemicals are dissolved, especially near the sites of former and present-day playas (dry lake beds) and lakes (Scarborough and Peirce, 1987). In addition, locally derived basin fill in the Colorado River valley interfingers with the late Miocene to Pliocene Bouse Formation, which accumulated in a brackish water estuary (salty, but less than sea water), probably linked to a proto-Gulf of California (Metzger and Loeltz, 1973; Lucchitta, 1979).

Based on drilling and geophysical data, the alluvial and basin-fill deposits are interpreted to be more than 600 m thick below the site (Figure 2.9). Basin fill is generally thickest near the center of a modern basin and thins gradually next to the flanking mountain

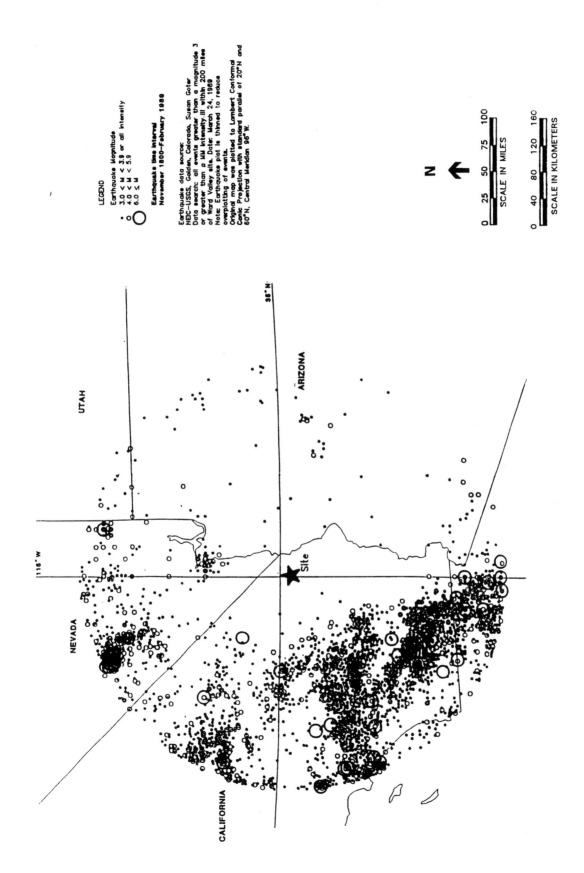

Figure 2.10   Seismic setting of Ward Valley - map of earthquakes in the region (from US Ecology, 1989).

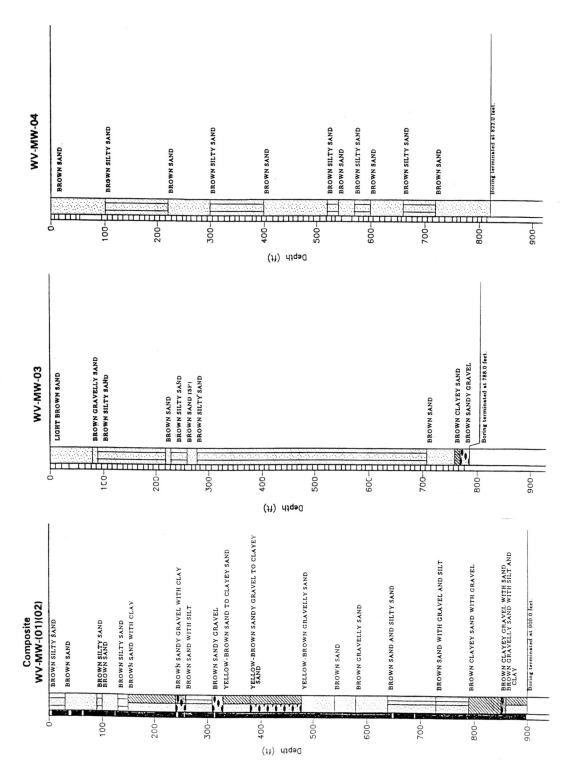

Figure 2.11 Well boring logs of alluvial sediments showing the character and variability of fan deposits. (Measurements originally taken in feet, not converted to metric). Locations of boreholes are found in Figure 2.4. (Harding Lawson Associates, 1991)

ranges. The basin fill takes the form of alluvial fans, which are gently sloping accumulations of intermittent stream deposits, and are common adjacent to mountain fronts. The fans may coalesce toward the basin, forming a bajada, a broad, low-relief feature that occurs at the proposed LLRW site.

The relative volumes of older (Miocene and Pliocene) basin fill versus Quaternary basin fill and alluvium is unknown. A fairly young age for some basin fill is indicated at Danby Dry Lake, where fossiliferous beds correlated with the Pliocene and upper Miocene Bouse Formation (Calzia, 1992) are overlain by 150 m of Pliocene and younger sediments (Wilshire et al., 1994).

The Ward Valley basin-fill deposits are interpreted to unconformably overlie highly faulted Miocene sedimentary and volcanic rocks that dip moderately to the west and that are in the upper plate of the regional detachment fault. A strong reflector, which is a surface deep within the crust that reflects acoustic or seismic waves, interpreted on a seismic reflection profile to be the detachment fault, dips gently to the east beneath the site at a depth of approximately 1.5 to 2 km. This detachment fault, when projected up-dip to the west, would probably reach the surface along the front of the Piute Mountains (Figure 2.9). The front of the Piute Mountains is interpreted to be the breakaway for the detachment, the place where the normal fault originally broke to the surface. Less faulted, more deep-level rocks, including metamorphic and granitic rocks similar to those in the Piute Mountains, are probably present beneath the detachment fault and its associated fractured and brecciated zone.

GEOMORPHOLOGY

The Ward Valley site and surroundings comprise three main geomorphic elements relevant to surficial processes and stability: (1) bounding mountain ranges trending approximately north - south, (2) gently sloping coalescing alluvial fan deposits on both sides of the valley forming bajadas, and (3) a mid-valley drainage (Homer Wash) with a north to south gradient. The local topographic features and geomorphology reflect the physiography of the Colorado River extensional corridor (Figure 2.1) and climate.

The Piute Mountains west of the site, where drainage across the alluvial fan originates, are the source of the sediments at the site surface. The Sacramento Mountains to the east of Homer Wash, across the valley from the Ward Valley site, shed sediments forming alluvial fan deposits east of Homer Wash.

The drainage channels in the vicinity of the site are very shallow and are interwoven into a braided network. Shallow channels indicate little incising or downcutting within the channels, and, in this area of low rainfall, all flow is ephemeral.

A single geomorphic surface covers much of the site area. The thin alluvial deposits on this surface are designated Qf1. Qf1 deposits show no soil development (LA Appendix 2310.A (Shlemon)), and, therefore, are inferred to be geologically young with immature soil development. The network of rills or shallow channels is developed on this surface in the vicinity of the site.

Local areas, slightly higher up the fan slope than the surface of Qf1, are underlain by moderately developed relict calcic soils (also commonly called caliche, see Box 2.1) or paleosols underlying a surface designated Qf2. These Qf2 paleosols also underlie much of the area under thin Qf1 deposits. From regional relationships, Qf2 paleosols are estimated to be 35,000 (35 ka) to 40,000 years (40 ka) and a deeper paleosol (Qf3), about 100 ka (LA Appendix 2310.A (Shlemon)). These calcic soils are consistent with limited rainfall for much of the period since formation.

The site features, i.e., surface deposits, paleosols, and the network of very shallow drainage channels indicate very little large-scale erosion or channelling under natural conditions for thousands of years. This observation is evidence for a relatively stable surface for the past thousands to tens of thousands of years.

Late Tertiary and Quaternary landscape development was strongly influenced by the local base level, the lowest level to which surface water flows within a drainage basin. Basins that are internally drained and unconnected with the regional drainage network or river system, such as Ward Valley, tend to be topographically higher and less deeply carved by surface water flow than basins of equivalent age along the Colorado River. Because the Colorado River is topographically lower, and the drop to the lower base level along the river is relatively rapid, it resulted in more erosional downcutting by rivers, removal of basin fill, and a lower overall elevation of the basin floor, for those basins that are connected with the regional drainage network leading to the Colorado. Drainages within the Colorado River trough carved deeply into the adjacent mountain ranges, commonly causing the ranges to be steepest and have their greatest local relief on the side facing the Colorado. The Stepladder Mountains and Sacramento Mountains display this topographic asymmetry.

The geomorphology and surface stability are discussed in more detail in Chapter 7 in relation to engineering considerations.

SUBSURFACE GEOPHYSICAL DATA

To characterize the site, geophysical methods have been used to constrain the depth and configuration of the boundary between basin fill and underlying bedrock and the depth to the water table.

The bedrock surface configuration at the proposed LLRW site is constrained by geologic, gravity, seismic-reflection, and ground-water data. This data set forms a basis for extrapolation and calibration of all data sets and also for extrapolation to the remainder of Ward Valley where the subsurface data are sparse. The geologic cross section (Figure 2.9) through the site effectively integrates these data sets into one interpretation of the geology across Ward Valley. Because of the nonuniqueness of these and all other geophysical and subsurface geologic data sets, however, other interpretations are possible.

The drill holes (LA Section 2320, Figure 2320 B–1), seismic reflection data, and electrical resistivity data (LA Section 2320, Figure 2320 B–6) provide useful information on the depth to the water table in the vicinity of the proposed LLRW site. These data also provide information on the possible existence of Tertiary ("old") alluvium and the geometry of

BOX 2.1

Background on "Caliche" or Calcic Soils

The term "caliche" has been, and is, commonly taken in southwestern United States to mean a calcareous or calcite-cemented unit developed by soil processes. However, this word has also been applied to other types of units. Currently, because of the broad and rather non-specific meaning the term has acquired, many geologists and soil scientists prefer the terms "calcic soil" or "pedogenic calcrete" as more precise (Machette, 1985). In addition to common soil unit identifiers, Shlemon (Appendix 2310.A) uses similar terms to describe the near-surface paleosols. It is clear from the context that, whatever the terminology used for the Ward Valley site, these units in the upper part of the sedimentary sequence are thought to be generally pedogenic in origin.

Soil scientists have developed specific nomenclature for the soil horizon developed in calcic soils (Gile and others, 1965). The "K horizon" is the master soil horizon, equivalent to more widely applied A, B, and C horizons. The K horizon has a very strong carbonate cemented fabric. Weaker calcic horizons are designated by k (e.g., Bk, as in LA Appendix 2310.A). Soils, including calcic soils, are the product of both environment and length of time over which they develop. The percentage of carbonate increases with depth in the soil and then decreases with depth still lower in the soil. The thickness of the subsurface soil horizons (e.g., B and K horizons), and the strength of individual soil properties used to describe the soil, increase with increasing age of the soil. Beneath progressively older surfaces the carbonates are progressively better developed. Calcic soils have been usefully classified in six stages (I-VI), from least to greatest morphological development, by Bachman and Machette (1977, p.40).

Within the administrative record, the near-surface calcic soils have been classified as stages II-III based on morphology. These stages commonly display a range of carbonate, from grain coatings and little matrix carbonate to continuous dispersion of carbonate in the matrix. At Stage II, clasts, or rock fragments, may show thin coatings of carbonate, and nodules develop where gravel content is low. By Stage III, cements between soil particles can be quite firm. The main hallmark differentiating these stages from Stage IV is the lack of platy to tabular structure within the K-fabric. Stages IV-VI show increasing induration, or hardness and platy, tabular, or laminar structure. Calcic soils at the Ward Valley site are not reported to show any of these more advanced morphologies.

Calcic soils develop with surface stability and appropriate climate. Because of the moderate solubility of calcium carbonate, greater rainfall and infiltration flushes carbonate components from the soil profile. In areas with higher rainfall, calcic soils rarely develop. Climatic changes, with increased rainfall, can destroy an existing calcic soil, and climatic fluctuations between arid and more humid environments can produce complex calcic soils or pedogenic calcretes such as the caprock on the Miocene Ogallala Formation exposed in eastern and southeastern New Mexico (Bachman and Machette, 1977). Calcic soils develop within the soil profile in a zone reflecting the mean annual depth of penetration by soil water (Bachman and Machette, 1977). The persistence of calcic soils at these depths indicates that the rainwater infrequently (or maybe never) reaches that level in large enough quantities to dissolve them.

Carbonate sources include soil gases from plant cover (e.g., Cerling and Quade, 1993), rain water, dust, and dissolved CO_2 in rainwater (Machette, 1985). Quade and others (1989) and Cerling and Quade (1993) demonstrate for calcic soils that soil carbon is generally equilibrated with soil gases characteristic of the plant cover and that oxygen in pedogenic carbonates is equilibrated with rainwater. Calcium sources include exposed older limestone formations, soil minerals, dissolved calcium in rainwater, and eolian dust.

Test pits at the Ward Valley site showed macromorphology and features interpreted by Shlemon (Appendix 2310.A) as soils or paleosols. In addition, some of the micromorphologies (pendulous calcite cement at 28.4 ft. depth, Figure 2310.A-11) are also indicative of pedogenic processes. The evidence presented within the administrative record is consistent with the presence of several calcic paleosols or pedogenic calcretes within the sequence. The calcic soils are appropriately classified as Stage II-III.

a number of bedrock structures. The seismic reflection records contain the best information available on the configuration of the bedrock surface, and form the primary basis, along with projection of surface geologic data, for suggesting that an irregular bedrock surface topography exists beneath the alluvial fill in Ward Valley.

Residual Bouguer gravity data for Ward Valley and vicinity, with the regional field removed (LA Section 2320, Figure 2320 B–12), provide a less precise means to extrapolate the configuration of the bedrock surface from the site throughout the area. Gravity data in the Basin and Range generally outline the configurations of basins and ranges, with highs mostly corresponding to ranges and lows mostly corresponding to basins. Using these observations and the detailed gravity profiles at the site and south of it in Ward Valley, two-dimensional models can be employed for interpretation of the bedrock surface topography in the northern two-thirds of Ward Valley (LA Section 2320, Figures 2320 B–9, B–14, B–15, B–16, and B–17). The bedrock surface configuration is portrayed as a narrow-floored, V-shaped canyon located west of the present-day Homer Wash drainage in a diagram by Harding Lawson Associates (Figure 2.12a). Based on these data sets, this is but one of several permissible interpretations of the available geologic, drilling, and geophysical data. Other permissible interpretations range from a simple flat-bottom valley to one interrupted by a tectonically derived fault-block topography (Figure 2.12).

Considering the residual Bouguer gravity profiles (particularly A–A' and D–D'; LA Section 2320, Figures 2320 B–9 and B–16) and the seismic reflection data (LA Section 2320, Figure 2320 B–6) in the site area, bedrock highs may exist down the axis of Ward Valley, but they probably do not intersect the present-day water table (Figure 2.13).

Vertical electrical soundings were performed using a Schlumberger array, as described in LA Appendix 2320.B, Geophysical Investigations. The general findings, with the exception of the two easternmost resistivity sites, were a comparatively high resistivity zone, approximately 100-200 ohm meters, extending to a depth of about 215 m, underlain by a lower resistivity (higher conductivity) zone of approximately 20-50 ohm meters. This resistivity boundary was interpreted as the water table, and its depth and identity were later verified by drilling at the site.

The two easternmost resistivity sites straddle Homer Wash, and in them the lower resistivities, 50-60 ohm meters, extend upward nearly to the surface. These lower resistivities cannot reasonably be interpreted as a mound on the water table, because, for example, the seismic reflection from the water table continues undeflected through this area (Appendix 2320 B). These lower resistivities near Homer Wash indicate some combination of higher pore water, greater silt and clay content, or higher salinity within the unsaturated zone. A decrease in sediment particle size (increase in silt and clay content) is expected toward Homer Wash, the axis of Ward Valley, and recharge to the unsaturated zone is also expected along Homer Wash. Clays, in particular, tend to lower resistivity by adsorbing electrolytes and water, and by increasing electrical connectivity.

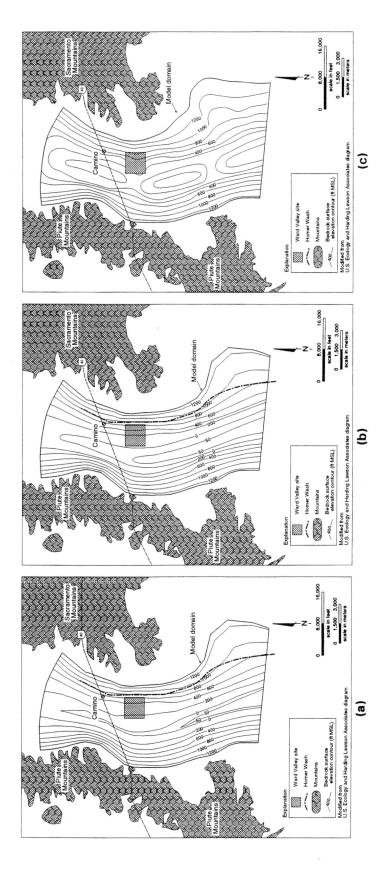

Figure 2.12 Bedrock surface topography in Ward Valley derived from gravity, seismic reflection, and other geophysical data. (a) Version presented in Harding Lawson's report (Eric Lappala presentation, August, 1994). (b) Version modified from (a) assuming a simple, broad, flat-bottom valley. (c) Further modification of (a) employing greater degree of interpretation of the residual Bouger gravity map to indicate possible fault-block modification of the bedrock surface paleotopography.

50

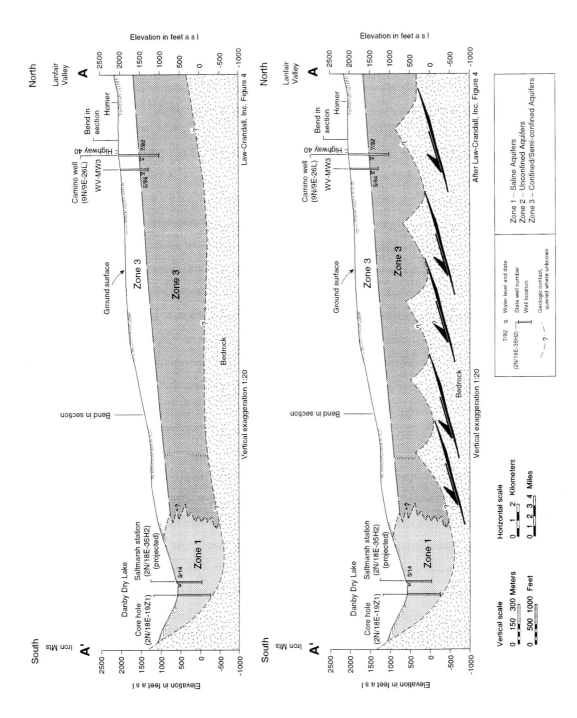

Figure 2.13 Longitudinal geologic sections through Ward Valley based on geologic, gravity, well and seismic-reflection data. (a) Interpretation presented in Law-Crandall, Inc. report (Figure 4). (b) Section modified from (a) assuming residual Bouger gravity to suggest that the axis of Ward Valley may consist of topography created by detachment faulting.

Climate

The climate of Ward Valley is typical of the eastern Mojave Desert. (Law/Crandall, 1992). Regional and local meteorological data reflect the climatic character of the Mojave Desert region for Ward Valley. Several sections of the administrative record provide general and specific data for the region and the Ward Valley site. There is no disagreement on the description of the general climatic characteristics of Ward Valley and the proposed site, though no long-term data from the site exist for engineering design.

The region is in a hot desert, with precipitation of 50 percent or less of potential evaporation and mean monthly air temperature of the coldest month greater than freezing. Winters are mild and summers are hot. The mean annual temperature is about 72°F (22°C) (Law/Crandall, 1992). Temperatures range from an extreme winter low of -2.4°C in January to extreme summer highs of about 49°C in July, with a wide daily temperature range (ca. 25°C) (LA Section 2200). Annual rainfall is estimated to be less than about 15 cm, and annual potential evapotranspiration is estimated between about 100 and 120 cm (LA Section 2200)[4]. Total annual pan evaporation based on five selected regional stations averages about 338 cm (LA Section 2200, Table 2200-6).

Some meteorological data were collected at the proposed radiological waste site from September 1987 through October 1988 (Harding Lawson Associates, 1994). For this one-year period, temperatures at a height above ground surface of 10 m ranged from 40.5°C to -3.3°C (LA Section 2200). At 0.5 m above the surface, the temperatures ranged from 43.9°C to -7.0°C. Rainfall data have been collected and reported at the site through August 1994 (Harding Lawson Associates, 1994b, 1994). For the 84-month period ending August 1994, monthly rainfall ranged from 0.00 cm (several months) to 11.5 cm (March 1992) and averaged 1.1 cm/month. The 84-month period yields an average equivalent of 13.4 cm of precipitation per year.

The Mojave Desert is characterized by low (10-15 cm), bimodal but winter-dominated precipitation (ca. 65 percent; Rowlands, 1980). A long-term, combined rainfall data record in the region taken at Needles and Needles airport for the 80-year period, 1903-1983 reported an average annual rainfall for this station of 11.8 cm (see footnote page 2-24) for this 80-year period (LA Section 2200, Table 2200-4, Law/Crandall, 1992). Rainfall data at Needles/Needles Airport for a 99-year period from 1892 to 1990 showed the maximum annual total rainfall for Needles was 34.1 cm (1939). Twenty-one out of the 99 years had total annual rainfall in excess of 15.2 cm (Law/Crandall, 1992).

Homer Wash watershed hydrology for the upper part of Ward Valley, where the LLRW facility is located, is described in Chapter 7.

---

[4] Note that rainfall figures vary slightly within this report depending on rainfall record location and period. Chapter 3 also uses rainfall to illustrate certain processes. The next paragraph gives more specific data from the Ward Valley site.

SURFACE-WATER HYDROLOGY

Homer Wash is the principal drainage feature in Ward Valley. It begins in the northern part of the valley and extends to Danby Dry Lake (LA Section 2410). The channel is more than 760 m east of the LLRW site at its nearest point. The alluvial fan-bajada surface has numerous shallow channels that feed drainage to Homer Wash (Figure 2.2 and 2.3).

Homer Wash is an ephemeral stream draining Ward Valley from north to south in a rather straight course. Homer Wash has little potential to develop sinuosity or extensive floodplain deposits for two reasons: (1) alluvial fan systems between these ranges will tend to maintain Homer Wash in a mid-valley position as long as there is no uneven tilting across the valley; (2) denser vegetation persists along the channel banks because ephemeral runoff is concentrated and may penetrate more deeply.

Hydrologic Soil Group Classification

Ward Valley has been characterized in terms of four hydrologic zones: (LA Section 2410):

Zone 1 is the outcropping rock and shallow soil areas formed by the Piute, Sacramento, Old Woman, Stepladder, and Turtle Mountains on the eastern and western boundaries of the basin. The San Bernardino County hydrologic soil group classification map for Ward Valley assigns soils in Zone 1 to the "D" or high runoff, low infiltration potential soils group. About 12 percent of the basin is classified as Zone 1. For hydrologic evaluation purposes, and based on reported visual inspection, vegetation in Zone 1 is classified as desert shrub with poor to fair cover.

Zone 2 represents upland area valleys between rock outcrop areas whose soils are formed by alluvial outwash materials that have migrated from the surrounding mountains. Approximately 25 percent of the basin is in Zone 2. Zone 2 soils are classified as "C", moderately high runoff or moderately low infiltration potential soils group. Vegetation in this zone is classified as desert shrub with poor to fair cover.

Zone 3 occurs on valley side slopes, where alluvial outwash from the mountain areas has formed alluvial fans that have coalesced into a bajada. The elevations range between 186 m and 914 m in this zone. Soils in Zone 3 are grouped into the "A" classification, and are characterized by high infiltration or low runoff rate. Type A soils chiefly consist of well to excessively drained sands and gravels. Zone 3 vegetation, for hydrologic evaluation purposes, is classified as desert shrub with fair to good cover. A unique surface feature of Zone 3 is the occurrence of isolated "hardpan" areas of less than .04 ha to 17 ha (0.1 acre to 42 acres). Hardpan is characterized by a hard surface, low infiltration, reduced vegetation density, and a slightly raised elevation above surrounding areas. Zone 3 represents about 60 percent of the basin.

Zone 4 exists only in the northern part of the Ward Valley Basin, where outcropping rock zones are not present on the basin boundary and different depositional processes from different source areas have occurred. Zone 4 soils fall into the "B" grouping, and are classified as having moderate runoff and infiltration rates. This soil group consists mainly of moderately well to well drained soils with moderately fine to moderately coarse textures. Vegetation in Zone 4 is hydrologically classified as fair to good desert shrub cover. Zone 4 comprises about 3 percent of the Ward Valley Basin (LA Section 2410).

HYDROGEOLOGY

Features

The hydrogeological setting of Ward Valley has many features in common with other basins in the Mojave Desert: (1) the valley is bordered on the east and west by mountain ranges; (2) bedrock in the valley is buried beneath a thick sequence of alluvial sediments; (3) the alluvial sediments, which are more than 600 m thick within Ward Valley, form the main aquifer system; (4) bedrock is generally considered to be much less permeable than the basin-fill alluvium.

Recharge

As there is little historic or present use of ground water within Ward Valley, the database to characterize the regional ground-water system is sparse. Precipitation falling in the mountains surrounding Ward Valley is the principal source of ground-water recharge. Some of the precipitation will enter deep bedrock flow systems through fractures in the bordering mountains. The rates of ground-water flow in the bedrock depend to a large extent on the degree to which fractures enhance the permeability of these crystalline rocks. Surface runoff from the mountains infiltrates on the upper reaches of the alluvial fans as the water encounters the more permeable alluvial sediments. If rainwater is able to infiltrate beneath the active root zone, the possibility exists for a general pattern of recharge across the floor of Ward Valley. Infrequent, high-precipitation events could lead to sporadic but volumetrically important recharge events centered beneath Homer Wash. Recharge may also enter northern Ward Valley as subsurface inflow from Lanfair Valley (Figure 2.1). Law/Crandall (1992) has estimated the annual recharge to the ground-water system in Ward Valley to be in the range from about 13.6 to 21.0 million cubic meters per year.

Water Table

The water table in Ward Valley is relatively deep; at the northern end of the valley it is up to 225 m beneath the ground surface. The extensive unsaturated zone above the water table and beneath the site of the proposed disposal facility is a consequence of this depth. At Danby Dry Lake, the local base level, or regional sink, for surface-water flow, the water table is located a few meters below the ground surface. The water table elevation ranges from about 437 m above sea level at Camino in the north to about 190 m above sea level at Danby Dry Lake. A ground-water divide is suspected to exist about 5 km north of the proposed facility (The Mark Group, 1990). North of this ground-water divide, ground water flows north and east into southern Piute Valley. South of this divide, ground water flows to the south through Ward Valley. The Danby Dry Lake playa is the main ground-water discharge area within Ward Valley. Law/Crandall (1992) estimate the regional water table gradient along Ward Valley to be 0.0037 m/m.

Yields of water wells in Ward Valley range from 38 to 1135 liters per minute (lpm) (Law/Crandall, 1992). The higher-yielding wells are located at the southern end of the valley and contain poor quality water. Wells at the northern end of the valley have low yields. Ground-water samples from the northern end of the valley suggest the ground water is a sodium bicarbonate type, with a total content of dissolved solids ranging between 300 and 500 milligrams per liter (mg/l). Fluoride, and in some cases nitrate, exceed drinking water standards.

ECOLOGY

Vegetation

Vegetation of the Mojave Desert has been divided into five subdivisions: creosote bush, shadscale, saltbush, blackbush, and Joshua tree (Vasek and Barbour, 1977). The Ward Valley site is located in the creosote bush subdivision. This subdivision is typically found in many of the broad valleys of the Mojave Desert with vegetation cover seldom over 30 percent. The most common association of this subdivision is creosote bush (*Larrea tridentata*) and burro bush or white bursage (*Ambrosia dumosa*) (MacMahon, 1988). Other shrubs locally common as associates of creosote bush and burro bush are spiny menodora (*Menodora spinescens*), wolfberries (*Lycium* spp.), Mormon tea (*Ephedra* sp.), ratany (*Krameria parvifolia*), goldenhead (*Acamptopappus schockleyi*), Fremont dalea (*Dalea fremontii*) and yellow paper daisy (*Psilostrophe cooperi*). Succulents such as cactus (e.g., beaver tail and chollas, *Opuntia* spp.; and barrel cactus, *Ferocactus acanthodes*) and yuccas (Mojave yucca, *Yucca schidigera*, and Joshua tree, *Y. brevifolia*) are commonly found within this desert but at low densities. Joshua tree may be an aspect dominant in some locations of the Mojave Desert but is not found on the Ward Valley site. Other shrubs, such as brittle bush (*Encelia farinosa*) and buckwheat (*Erigonum fasciculatum*), are also represented but seldom with high densities except in local microenvironments, often early successional or disturbance sites. Herbaceous plants within the Mojave Desert and in Ward Valley are primarily annuals, comprising about 60 percent of the flora and often producing dense cover following winter or summer rain events. Perennial grasses such as *Hilaria* and *Stipa* have been found on the site.

Xeroriparian vegetation, which refers to species typically found in or along washes in arid environments, is not well developed along Homer Wash, with surface flows following only heavy precipitation events. Xeroriparian species found along the wash include catclaw acacia (*Acacia greggii*) and smoke tree (*Dalea spinosa*). Creosote bush, brickel bush

(Brickellia incana), cheese bush *(Hymenoclea salsola)*, Anderson's thornbush *(Lycium andersonii)*, and punctate rabbit-brush *(Chrysothamnus paniculatus)* may also occur along the wash. Facultative xeroriparian species are found along large washes in the Mojave Desert as well as being associated with rills and small channels.

Desert Fauna

Fauna of Ward Valley are also typical of the Mojave Desert. Vertebrates include predatory species such as coyote, and birds of prey (e.g., hawks and ravens); other birds such as Gambel's quail and mourning dove; rodents such as kangaroo rats, pocket mice, and antelope ground squirrels; and reptiles such as chuckwalla and desert tortoise. Like most deserts, the Mojave supports a wide variety of invertebrates, for example, spiders, ants, and termites.

Several studies document the diversity of the California deserts which includes the Mojave. Bradley and Deacon (1967) describe the biotic communities of southern Nevada, an area just north of the Ward Valley site. They reported thirty species of reptiles, forty bird species, and forty-four species of mammals. Laudenslayer and Boyer (1980), studying mammals of the California deserts, reported ninety-four mammalian species. The order Rodentia comprised more than half of these species, including 21 species of bats. England and Laudenslayer et al. (1980) reported 427 species of birds in the California deserts.

Critical Habitat

In 1989, the U.S. Fish and Wildlife Service (USFWS) listed the Mojave desert tortoise as an endangered species in an emergency rule, later revised in a final rule dated April 2, 1990 (USFWS, 1989; USFWS, 1990) as a threatened species. Subsequently, in 1994, USFWS designated a large area of the eastern Mojave Desert as critical habitat for the desert tortoise (USFWS, 1994a). The 6.4 million acres identified included Ward Valley. A Recovery Plan, published in June 1994, identified a number of Desert Wildlife Management Areas (DWMA) within the critical habitat to implement recovery actions that were designed to assure the continued existence of the desert tortoise and of the ecosystem upon which the tortoise depends (USFWS, 1994b). Ward Valley is in the Chemehuevi DWMA. The density of the tortoise in the Chemehuevi DWMA is listed as 10-275 adults per square mile (USFWS, 1994b), which is among the higher concentrations of Mojave Desert tortoise in the 14 DWMAs. In addition, the degree of threat in the Chemehuevi DWMA is rated as one, which is the low end of a scale from one to five. Only two of the 14 DWMAs are considered to be areas where the threat to the desert tortoise is this low (USFWS, 1994b).

REFERENCES

Bachman, G. O., and M. N. Machette. 1977. Calcic soils and calcretes in the southwestern United States. Open File Report 77-794. U.S. Geological Survey. Washington, D. C., 163 p.

Bender, E. E. 1990. Geology of the Fenner Gneiss, Piute and Old Woman Mountains, San Bernardino County, California. M.S. thesis. Vanderbilt University, Nashville. 161 pp.

Bohannan, R. G. 1984. Nonmarine sedimentary rocks of Tertiary age in the Lake Mead region, southeastern Nevada and northwestern Arizona. U.S. Geological Survey Professional Paper 1259. 72 pp.

Bradley, W. G. and J. F. Deacon. 1967. The biotic communities of southern Nevada. Nevada State Museum Anthropological Papers 13.

Bryant, B., C. M. Conway, J. E. Spencer, S. J. Reynolds, J. O. Otton, and P. M. Blacet. 1992. Geologic map and cross section across the boundary between the Colorado Plateau Transition Zone and the Basin and Range southeast of Bagdad Arizona. U.S. Geological Survey Open-File Report 92(428):23, 2 plates, scale 1:100,000.

Calzia, J. P. 1992. Geology and saline resources of Danby Dry Lake playa, southeastern California, *in* Old Routes to the Colorado, Reynolds, R. E., compiler. San Bernardino County Museum Association special publication 92(2):87-91.

Carr, W. J., and D. D. Dickey. 1980. Geologic map of the Vidal, California, and Parker SW, California-Arizona quadrangles. U.S. Geological Survey Miscellaneous Investigations Series Map I-1125, scale 1:24,000.

Carr, W. J., D. D. Dickey, and W. D. Quinlivan. 1980. Geologic map of the Vidal NW, Vidal Junction and parts of the Savahia Peak SW and Savahia Peak quadrangles. San Bernardino County, California: U.S. Geological Survey Miscellaneous Investigations Map I-1126, scale 1:24,000.

Carr, W. J. 1991. A contribution to the structural history of the Vidal-Parker region, California and Arizona. U.S. Geological Survey Professional Paper 1430. 40 pp.

Cerling, T. E. and J. Quade. 1993. Stable carbon and oxygen isotopes in soil carbonates. in Continental Isotopic Records, Geophysical Monograph 78. American Geophysical Union. pp. 217-231

Collier, J. T. 1960. Geology and mineral resources of Township 8 North, Ranges 19 and 20 East, San Bernardino base and meridian. San Bernardino County, California. San Francisco, Southern Pacific Land Company, scale 1:24,000.

Custis, K. H. 1984. Geology and dike swarms of the Homer Mountain area, San Bernardino County, California. M.S. thesis. California State University at Northridge. 168 pp.

Davis, G. A., J. L Anderson, E. G. Frost, and T. J. Shackelford. 1980. Mylonitization and detachment faulting in the Whipple-Buckskin-Rawhide Mountains terrain, southeastern California and western Arizona, *in* Cordilleran Metamorphic Core Complexes, Crittenden, M. D., Jr., P. J. Coney, and G. H. Davis, eds. Geological Society of America Memoir 153:79-129.

Davis, G. A., G. S. Lister, and S. J. Reynolds. 1986. Structural evolution of the Whipple and South Mountains shear zones, southwestern United States. Geology 14:7-10.

Davis, G. A. 1988. Rapid upward transport of mid-crustal mylonitic gneisses in the footwall of a Miocene detachment fault, Whipple Mountains, southeastern California. Geologische Rundschau 77:191-209.

Davis, G. A. and G. S. Lister. 1988. Detachment faulting in continental extension; Perspectives from the southwestern U.S. Cordillera. Geological Society of America Special Paper 218:133-159.

Demsey, K. A. 1988. Geologic map of Quaternary and Upper Tertiary alluvium in the Phoenix North 30' x 60' quadrangle, Arizona. Arizona Geological Survey Open-File Report 88-17, scale 1:100,000.

Dickey, D. D., W. J. Carr and W. B. Bull. 1980. Geologic map of the Parker NW, Parker, and parts of the Whipple Mountains SW and Whipple Wash quadrangles, California and Arizona. U.S. Geological Survey Miscellaneous Investigations Map I-1124, scale 1:24,000.

Fedo, C. M. and J. M. G. Miller. 1992. Evolution of a Miocene half-graben basin, Colorado River extensional corridor, southeastern California. Geological Society of America Bulletin 104:481-493.

Fletcher, J. M. and K. E. Karlstrom. 1990. Late Cretaceous ductile deformation, metamorphism, and plutonism in the Piute Mountains, eastern Mojave Desert. Journal of Geophysical Research 95:487-500.

Foster, D. A, T. M. Harrison, and C. F. Miller. 1989. Age, inheritance, and uplift of the Old Woman-Piute batholith, California, and implications for K-feldspar age spectra. Journal of Geology 97:232-243.

Foster, D. A., T. M. Harrison, C. F. Miller, C. F., and K. A. Howard. 1990. The $^{40}Ar/^{39}Ar$ thermochronology of the eastern Mojave Desert, California, and adjacent Arizona with implications for the evolution of metamorphic core complexes. Journal of Geophysical Research 95:20,005-20,024.

Foster, D. A., C. F. Miller, T. M. Harrison, and T. D. Hoisch. 1992. $^{40}Ar/^{39}Ar$ thermochronology and thermobarometry of metamorphism, plutonism, and tectonic denudation in the Old Woman Mountains area, California. Geological Society of America Bulletin 104:176-191.

Foster, D. A., D. S. Miller, and C. F. Miller. 1991. Tertiary extension in the Old Woman Mountains area, California: Evidence from apatite fission track analysis. Tectonics 10:875-886.

Frost, E. G. and D. A. Okaya. 1986. Application of seismic reflection profiles to tectonic analysis in mineral exploration, *in* Frontiers in Geology and Ore Deposits of Arizona and the Southwest, Beatty, B. and P. A. K. Wilkinson, eds. Arizona Geological Society Digest 16:137-152.

Gile, L. H., P. F. Peterson, and R. B. Grossman. 1965. The K horizon - master soil horizon of carbonate accumulation. Soil Science, 99(2):74-82.

Harding Lawson Associates. 1994. Letter Report to Ms. Ina Alterman; Supplemental information regarding tritium vapor sampling. Dated October 12, 1994.

Hileman, G. E., C. F. Miller, and M. A. Knoll. 1990. Mid-Tertiary structural evolution of the Old Woman Mountains area: Implications for crustal extension across southeastern California. Journal of Geophysical Research 95:581-597.

Hillhouse, J. W., and R. E. Wells. 1991. Magnetic fabric, flow directions, and source area of the lower Miocene Peach Springs Tuff in Arizona, California, and Nevada. Journal of Geophysical Research 96:12,443-12,460.

Hoisch, T. D., C. F. Miller, M. T. Heizler, T. M. Harrison, and E. F. Stoddard. 1988. Late Cretaceous regional metamorphism in southeastern California. Pp. 539-571 *in* Metamorphism and Crustal Evolution of the Western United States-Rubey Volume VII, Ernst, W. G., ed. New Jersey: Prentice Hall.

Howard K. A., B. E. John and C. F. Miller. 1987. Metamorphic core complexes, Mesozoic ductile thrusts, and Cenozoic detachments: Old Woman Mountains - Chemehuevi Mountains transect, California and Arizona, *in* Geologic Diversity of Arizona and Its Margins: Excursions to Choice Areas, Davis, G.H., and E.M. Vandendolder, eds. Arizona Bureau of Geology and Mineral Technology Special Paper 5:365-382.

Howard, K. A., and B. E. John. 1987. Crustal extension along a rooted system of imbricate low-angle faults: Colorado River extensional corridor, California and Arizona, *in* Continental Extensional Tectonics, Coward, M. P., J. F. Dewey, and P. L. Hancock, eds.. Geological Society of London Special Publication 28: 299-311.

Howard, K. A., P. Stone, M. A. Pernokas, and R. F. Marvin. 1982. Geologic and geochronologic reconnaissance of the Turtle Mountains area, California: West border of the Whipple Mountains detachment terrain, *in* Mesozoic-Cenozoic tectonic Evolution of the Colorado River Region, California, Arizona, and Nevada (Anderson-Hamilton Volume), Frost, E. G. and D. L. Martin, eds. San Diego: Cordilleran Publishers, pp. 341-355.

John, B. E. and K. A. Howard. 1994. Drape folds in the highly attenuated Colorado River extensional corridor, California and Arizona, *in* Geological Investigations of an Active Margin, McGill, S. F. and T. M. Ross, eds. Geological Society of America Cordilleran Section Guidebook, 27th Annual Meeting, San Bernardino, California, March 21-23, 1994. San Bernardino County Museum Association, pp. 94-106.

John, B. E. 1987. Geometry and evolution of a mid-crustal extensional fault system: Chemehuevi Mountains, southeastern California, *in* Continental Extensional Tectonics, Coward, M. P., J. F. Dewey, and P. L. Hancock, eds. Geological Society of London Special Paper 28, pp. 313-335.

John, B. E. and D. A. Foster. 1993. Structural and thermal constraints on the initiation angle of detachment faulting in the southern Basin Range: The Chemehuevi Mountains case study. Geological Society of America Bulletin 105:1091-1108.

Karlstrom, K. E., C. F. Miller, J. A. Kingsbury and J. L. Wooden. 1993. Pluton emplacement along an active ductile thrust zone, Piute Mountains, southeastern California: Interaction between deformational and solidification processes. Geological Society of America Bulletin 105:213-230.

Laudenslayer, W. F. Jr. and K. B. Boyer. 1980. Mammals of the California Desert, *in* The California Desert: An Introduction to Natural Resources and Man's Impact, J. Latting, ed. California Native Plant Society Special Publication, 5.

Laudenslayer, W. F. Jr., S. W. Cardiff, and A. S. England. 1980. Checklist of birds known to occur in the California Desert, *in* The California Desert: An Introduction to Natural Rresources and Man's Impact, J. Latting, ed. California Native Plant Society Special Publication, 5.

Law/Crandall, Inc. August 19, 1992. Water Resources Evaluation, Ward Valley, San Bernardino County, California, prepared for City of Needles.

License Application. 1989. U.S. Ecology, Inc. Administrative Record, Ward Valley Low-Level Radioactive Waste Disposal Facility, Sections 2110, 2200, 2410, Appendices 2310.A (Shlemon), 2320.B.

Lucchitta, I. 1972. Early history of the Colorado River in the Basin and Range province. Geological Society of America Bulletin 83:1933-1947.

Lucchitta, I. 1974. Structural evolution of northwest Arizona and its relation to adjacent Basin and Range Province structures, *in* Geology of Northern Arizona, with notes on Archaeology and Paleoclimate, pt. 1, regional studies, Karlstrom, T. N. V., G. A. Swan, R. L. Eastwood, eds. Papers from the 27th annual meeting of the Geological Society of America Rocky Mountain Section. Flagstaff.

Machette, M. N. 1985. Calcic soils of the southwestern United States *in* Soils and Quaternary Geology of the Southwestern United States, Weide, D. L., ed. Special Paper 203, Geological Society of America. pp. 1-21

Menges, C. M. and P. A. Pearthree. 1989. Late Cenozoic tectonism in Arizona and its impact on regional landscape evolution, *in* Geologic Evolution of Arizona, Jenney, J. P. and S. J. Reynolds, eds. Arizona Geological Society Digest 17:649-680.

MacMahon, J. 1988. Warm deserts *in* North American Vegetation, M. Barbour and W.D. Billings, eds. New York: Cambridge University Press. pp. 231-264

McCarthy, J., S. P. Larkin, G. S. Fuis, R. W. Simpson, and K. A. Howard. 1991. Anatomy of a metamorphic core complex: Seismic refraction wide-angle reflection profiling *in* Southeastern California and Western Arizona. Journal Geophysical Research 96:12, 259-12, 291.

McClelland, W. C. 1982. Structural geology of the central Sacramento Mountains, San Bernardino County, California, *in* Mesozoic-Cenozoic Tectonic Evolution of the Colorado River Region, California, Arizona, and Nevada (Anderson-Hamilton Volume), Frost, E. G. and D. L. Martin, eds. San Diego: Cordilleran Publishers, pp. 401-407.

Metzger, D. G. and O. J. Loeltz. 1973. Geohydrology of the Needles area, Arizona, California, and Nevada. U.S. Geol. Survey Prof. Paper 486-J. 54 pp.

Miller, C. F., J. M. Hanchar, J. L. Wooden, V. C. Bennet, T. M. Harrison, D. A. Wark, and D. A. Foster. 1992. Source region of a granite batholith: evidence from lower crustal xenoliths and inherited accessory minerals. Geological Society of America Special Paper 272, pp. 49-62.

Miller, C. F., K. A. Howard and T. D. Hoisch. 1982. Mesozoic thrusting, metamorphism, and plutonism, Old Woman-Piute Range, southeastern California. Pp. 561-581 *in* Mesozoic-Cenozoic Tectonic Evolution of the Colorado River Region, California, Arizona, and Nevada (Anderson-Hamilton Volume), Frost, E. G. and D. L. Martin, eds. San Diego: Cordilleran Publishers.

Quade, J., T. E. Cerling, and J. R. Bowman. 1989. Systematic variation in the stable carbon and oxygen isotopic composition of pedogenic carbonate along elevation transects *in* the Southern Great Basin, USA. Geological Society of America Bulletin, 101:464-475.

Reynolds, S. J. 1988. Geologic Map of Arizona: Arizona Geological Survey Map 26, scale 1:1,000,000.

Reynolds, S. J., and G. S. Lister. 1987. Structural aspects of fluid-rock interactions in detachment zones. Geology 15:362-366.

Reynolds, S. J., and G. S. Lister. 1990. Folding of mylonitic zones in Cordilleran metamorphic core complexes - Evidence from near the mylonitic front. Geology 18:216-219.

Reynolds, S. J., F. P. Florence, J. W. Welty, M. S. Roddy, D. A. Currier, A. V. Anderson, and S. B. Keith. 1986. Compilation of radiometric age determinations in Arizona. Arizona Bureau of Geology and Mineral Technology Bulletin 197:1-258.

Reynolds, S. J., and W. A. Rehrig. 1980. Mid-Tertiary plutonism and mylonitization, South Mountains, central Arizona, in Cordilleran metamorphic core complexes, Crittenden, M. D., Coney, P. J., and Davis, G. H., eds. Geological Society of America Memoir 153:159-174.

Reynolds, S. J., and J. E. Spencer. 1985. Evidence for large-scale transport on the Bullard detachment fault, west-central Arizona. Geology 13:353- 356.

Reynolds, S. J., S. M. Richard, G. M. Haxel, R. M. Tosdal, and S. E. Laubach. 1988. Geologic setting of Mesozoic and Cenozoic metamorphism in Arizona. *in* Metamorphism and Crustal Evolution of the Western United States, Ernst, W.G., ed. Englewood Cliffs: Prentice- Hall. pp. 466-501

Reynolds, S. J., M. Shafiqullah, P. E. Damon, and E. DeWitt. 1986. Early Miocene mylonitization and detachment faulting, South Mountains, central Arizona. Geology 14:283-286.

Reynolds, S. J., J. E. Spencer, S. M. Richard, and S. E. Laubach. 1986. Mesozoic structures in west-central Arizona, *in* Frontiers in Geology and Ore Deposits of Arizona and the Southwest, Beatty, B. and Wilkinson, P. A. K., eds. Arizona Geological Society Digest 16:35-51.

Robinson, F. W. 1988. Petrology of Goffs pluton, northern Old Woman-Piute Range, southeastern California. M.S. thesis. Vanderbilt University, Nashville. 104 pp.

Roddy, M. S., S. J. Reynolds, B. M. Smith, and J. Ruiz. 1988. K-metasomatism and detachment-related mineralization, Harcuvar Mountains, Arizona. Geological Society of America Bulletin 100:1627-1639.

Rowlands, P. 1980. The vegetation attributes of the California deserts, The California Desert: An Introduction to Its Resources and Man's Impact, J. Latting, ed. California Native Plant Society Special Publication 5.

Rowlands, P., H. Johnson, E. Riter, and A. Endo. 1982. The Mojave Desert. *in* Reference Handbook on the Deserts of North America, G.L. Bender, ed. Westport, CN:Greenwood Press. Pp. 103-145.

Scarborough, R. B., and H. W. Peirce. 1978. Late Cenozoic basins of Arizona, *in* Land of Cochise. New Mexico Geological Society 29th Field Conference Guidebook, J.F. Callender, J.C. Wilt, and R.E. Clemons, eds. pp. 253-259.

Shafiqullah, M., P. E. Damon, D. J. Lynch, S. J. Reynolds, W. A. Rehrig, and R. H. Raymond. 1980. K-Ar geochronology and geologic history of southwestern Arizona and adjacent areas *in* Studies in Western Arizona, J. P. Jenney, and C. Stone, eds. Arizona Geological Society Digest 12. Pp. 202-260.

Sherrod, D. R. and J. E. Nielson, eds. 1993. Tertiary stratigraphy of highly extended terranes, California, Arizona, and Nevada. U.S. Geological Survey Bulletin 2053. 250 pp.

Simpson, C., J. Schweitzer, and K. A. Howard. 1991. A reinterpretation of the timing, position, and significance of the Sacramento Mountains detachment fault, south eastern California. Geological Society America Bulletin 103:751-761.

Smith, B. M., S. J. Reynolds, H. W. Day, and R. Bodnar. 1991. Deep-seated fluid involvement in ductile-brittle deformation and mineralization, South Mountains metamorphic core complex, Arizona. Geological Society of America Bulletin 103:559-569.

Spencer, J. E., and S. J. Reynolds. 1989. Middle Tertiary tectonics of Arizona and adjacent areas *in* Geologic Evolution of Arizona, J. P. Jenney, and S. J. Reynolds, eds. Arizona Geological Society Digest 17. Pp. 539-573.

Spencer, J. E., and S. J. Reynolds. 1990. Relationship between Mesozoic and Cenozoic tectonic features in west-central Arizona and adjacent southeastern California. Journal of Geophysical Research 95:539-555.

Spencer, J. E., and S. J. Reynolds, eds. 1989. Geology and mineral resources of the Buckskin and Rawhide Mountains, west-central Arizona. Arizona Geological Survey Bulletin 198. 280 pp.

Spencer, J. E., and J. W. Welty. 1989. Mid-Tertiary ore deposits in Arizona, *in* Geologic Evolution of Arizona, Jenney, J. P., and Reynolds, S.J., eds. Arizona Geological Society Digest 17. Pp. 585-608.

Spencer, J. E. 1985. Miocene low-angle normal faulting and dike emplacement, Homer Mountain and surrounding areas, southeastern California and southernmost Nevada. Geological Society of America Bulletin 96:1140-1155.

Spencer, J. E. and R. D. Turner. 1983. Geologic map of the northwestern Sacramento Mountains, southeastern California. U.S.Geological Survey Open-file Report 83-614, scale 1:24,000.

Spencer, J. E. and R. D. Turner. 1985. Geologic map of Homer Mountain and the southern Piute Range, southeastern California. U.S. Geological Survey Miscellaneous Field Studies Map MF-1709, scale 1:24,000.

Spurk, W. H. 1960. Geology and mineral resources of Township 9 North, Ranges 19 and 20 East, San Bernardino base and meridian, San Bernardino County, California. San Francisco, Southern Pacific Land Company, scale 1:24,000.

Staude, J. M. G. and C. F. Miller. 1987. Lower crustal xenoliths from a Tertiary composite dike, Piute Mountains, S.E. California. Geological Society of America Abstracts with Programs 19:454.

Stone, P., K. A. Howard and W. Hamilton. 1983. Correlation of Paleozoic strata of the southeastern Mojave Desert region, California and Arizona. Geological Society of America Bulletin 94:1135-1147.

Tischler, M. S. 1960. Geology and mineral resources of Township 10 North, Ranges 19 and 20 East, San Bernardino base and meridian, San Bernardino County, California. San Francisco, Southern Pacific Land Company, scale 1:24,000.

U.S. Fish and Wildlife Service. 1989. Endangered and threatened wildlife and plants; emergency determination of endangered status for the Mojave population of the desert tortoise. Federal Register 54(149):32326.

_____. 1990. Endangered and threatened wildlife and plants; determination of threatened status for the Mojave population of the desert tortoise. Federal Register 55(63):12178-91.

_____. 1994. Endangered and threatened wildlife and plants; determination of critical habitat for the Mojave population of the desert tortoise. Federal Register 59(26):5820-66.

U.S. Geological Survey. 1984. West of Flattop Mountain Quadrangle, California - San Bernardino County, 7.5 Minute Series (Topographic). Washington, D.C.: U.S. Department of the Interior.

Vasek, F. C., and M. G. Barbour. 1977. Mojave desert scrub vegetation. Pp. 835-867 in Terrestrial vegetation of California, M. G. Barbour and J. Major, eds. New York: Wiley.

Wernicke, B., and B. C. Burchfiel. 1982. Modes of extension tectonics: Journal of Structural Geology 4:105-115.

Wernicke, B. 1981. Low-angle normal faults in the Basin and Range Province: Nappe tectonics in an extending orogen. Nature 291:645-648.

Wernicke, B. 1985. Uniform-sense normal simple shear of the continental lithosphere. Canadian Journal of Earth Sciences 22:108-125.

Wilshire, H. G., K. A. Howard, and D. M. Miller. 1994. Ward Valley proposed low level radioactive waste site. A report to the National Academy of Sciences.

Wooden, J. L. and D. M. Miller. 1990. Chronologic and isotopic framework for Early Proterozoic crustal evolution in the eastern Mojave Desert Region, SE California. Journal of Geophysical Research 95:20,133-20,146.

3

RECHARGE THROUGH THE UNSATURATED ZONE

Issue 1[1]
Potential Transfer of Contaminants
Through the Unsaturated Zone to the Ground Water

THE WILSHIRE GROUP POSITION

The Wilshire group questioned the adequacy of the evaluation in the license application and supporting documents of the nature of water movement in the thick unsaturated zone beneath the Ward Valley site (Wilshire et al, 1993a, 1993b). They divided this issue into five subissues. These are:

1. The soil properties measured and used in the modeling of water movement in the unsaturated zone do not adequately represent the variability and complexity of the materials present.
2. No consideration was given to rapid migration of water along preferential pathways.
3. The reporting of tritium at depths of up to 30 m indicates vertical water transport at rates much faster than that used in the performance-assessment models.
4. Electrical sounding data from along Homer Wash suggest the possible existence of higher water content that may indicate ground-water recharge from the wash.
5. Interpretation of both stable and radioactive tracers found in the ground water in the license application was incorrect and could be interpreted as evidence of recent recharge through the unsaturated zone.

THE DHS/U.S. ECOLOGY POSITION

The DHS concluded that, based on the general aridity of the site, hydraulic data collected from boreholes, the interpreted age of the ground water, and a simplified tritium diffusion model, the site shows no evidence for high water flux or rapid pathways.

---

[1] In this report, the Wilshire group Issues 1 and 2 have been reversed. Issue 1 in this report is actually Issue 2 of the Wilshire group and vice versa. The committee found it more useful to address the Wilshire group's Issue 2 on the potential for recharge through the unsaturated zone first because it requires extensive discussion of the properties and characteristics of the unsaturated zone. Issue 1 of the Wilshire group, the potential for later flow in the shallow subsurface (Issue 2 of this report), requires much less discussion of unsaturated zone characteristics and therefore draws upon the information in the preceding chapter.

THE COMMITTEE'S APPROACH

The committee grouped the subissues raised by Wilshire et al. (1993b) into three broad questions relevant to the behavior of water in the unsaturated zone:

1. What is the nature, direction, and magnitude of soil-water movement beneath the Ward Valley site, including the influence of geologic heterogeneity and the relative importance of piston and preferential flow (Wilshire et al., 1993a, b; subissues 2, 3, and 4)?

2. Are the data and models used to quantify the rate and direction of water and contaminant movement through the unsaturated zone adequate to define the general performance characteristics (Wilshire et al., 1993b, subissue 1)?

3. Does ground water below the Ward Valley site show evidence of recent recharge (Wilshire et al., 1993a, b; subissue 5)?

In (1), the discussion of the broad issue of soil-water movement will address the three subissues. For each of these questions, the committee has looked closely at the data available from both the site license and supporting documents, at the relevant scientific literature, as well as at documents supplied by interested individuals and organizations at the open meetings and subsequently.

To answer each of these questions, a brief review of the nature of soil-water movement and its implications for waste disposal will put the issues in perspective. Data from previous studies of water movement in the unsaturated zone with emphasis on arid regions are instructive. These previous studies show that the science of unsaturated-zone hydrology is young and that the techniques employed continue to have significant uncertainties. As a result, the committee considered it prudent to base its conclusions on the results of multiple and independent lines of evidence, not only of the most likely estimate of water velocity, but also to assess the performance impacts of other, less likely, estimates.

THE NATURE OF WATER MOVEMENT IN THE UNSATURATED ZONE

In a typical arid setting, most precipitation infiltrates the soil, except under conditions of intense precipitation. Some of this water evaporates and some is transpired by plants. Together these two processes, which return rain water to the atmosphere, are termed evapotranspiration. If the volume infiltrated is greater than that evapotranspired, it can result in deep downward or lateral percolation. If the unsaturated zone is thin or the percolation rate high, the recharge rate at the water table will approach the rate of percolation over a long time. If, however, the unsaturated zone is very deep and the percolation rate is small, then the recharge rate may be out of time-phase with the current near-surface conditions. In such cases, it is important to differentiate between measurements of deep percolation in the unsaturated zone and estimates of recharge made at the water table.

Water Content

Water content is a standard soil physics parameter useful for evaluating water movement in the unsaturated zone is water content. Water content is either measured in the laboratory using soil samples collected in the field or monitored in the field using a neutron probe or time-domain reflectometry to determine the soil dryness.

Water-content monitoring can be used to evaluate the movement of water pulses through the unsaturated zone. In general, errors associated with the typical water-content measurement techniques (neutron probe or time-domain reflectometry) are generally ±1 percent volumetric water content. This may not be sufficiently accurate to detect small fluxes that could move through the deep unsaturated zone of desert soils. Under conditions of steady flow, water content will not vary; therefore, the absence of temporal variations in water content does not necessarily preclude downward flow. In such a case, the flow is controlled by the hydraulic conductivity of the soil. Water-content monitoring is most applicable for measuring large subsurface water fluxes under transient conditions. In addition, water content is discontinuous across different soil types; therefore, variations in water content with depth in heterogeneous soils do not necessarily indicate the direction of water movement.

Water content data are available for many sites from interstream settings and show that water contents in desert soils are generally low. Long-term monitoring of soil water has been used to evaluate infiltration and deep percolation in several arid sites. Some of these studies were carried out at a site in the Chihuahuan Desert of New Mexico. Rainfall at the New Mexico site is approximately twice that at Ward Valley. The seasonal distribution of rainfall at the New Mexico site also differs from that at Ward Valley, with more summer precipitation at the New Mexico site, which is typical of the Chihuahuan Desert, and more winter precipitation at the Ward Valley site, typical of the Mojave Desert. Winter precipitation is more effective at infiltrating the soil than summer precipitation because of lower evapotranspiration in the winter. Long term precipitation records (1953 - 1989) at the Jornada Experiment Range in New Mexico indicate that only 23 percent of the rain falls between November and March whereas 50 percent of the rain falls between November and March in Needles, California (period of record 1948 to 1989). Although the seasonal distribution of rainfall differs between the two sites, the amount of winter rainfall is similar (5.6 cm Needles; 5.9 cm Jornada Experiment Station); therefore, results of studies at the New Mexico site are applicable to the Ward Valley site.

Long-term studies in the Chihuahuan Desert measured soil moisture along a transect from an ephemeral lake through the piedmont and up a steep rock slope (Wierenga et al., 1987). Large spatial variability in water content was recorded from a mean of 0.3 m^3/m^3 (± 0.04 m^3/m^3) at 130 cm depth in the playa to 0.04 m^3/m^3 (± 0.04 m^3/m^3) at the upper rocky end of the transect (Figure 3.1). This variability reflects in part the textural variations from the clayey soils in the playa to more gravelly soils at the rocky slope. Temporal variations in water content were measured for a site near the center of the transect. This site is located on an alluvial fan with a deep loamy soil and is vegetated with creosote bush similar

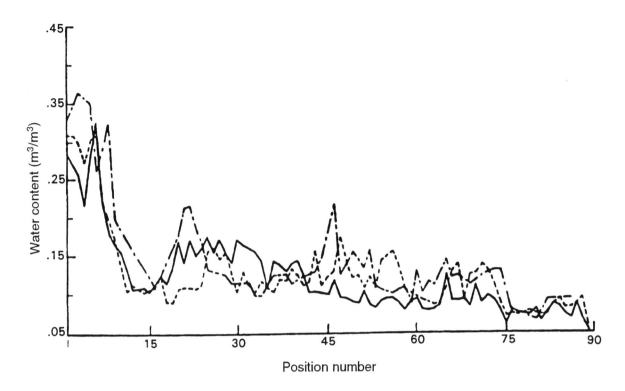

Figure 3.1   Variation in water content on an alluvial fan along a 3 km transect in the Jornada Range in southern New Mexico. Water contents presented for depths of 30, 90, 130 cm during the week of 30 April - 6 May 1992. Symbols: —, 30 cm; ----, 90 cm; –•–, 130 cm. Vegetation is creosote bush and annual rainfall is 20 cm.

to the Ward Valley site. Although the soils at the Las Cruces site in New Mexico and at Ward Valley are similar in terms of their soil hydraulic properties, we recognize that the ecological systems at the two sites are different.

The data at the Las Cruces site display significant variations in water content at 30 cm depth during six of the nine years studied. At a depth of 130 cm, fluctuations in water content occurred only in response to the relatively wet fall and winter of 1985. Thus during eight of the nine years, rainfall that infiltrated the soil was taken up by the plant roots in the upper 130 cm of the root zone and evapotranspired back into the atmosphere. Plant roots were observed in a 6 m deep trench near the trench down to 4 m depth. Most likely, water that passed the 130 cm depth was taken up by plant roots below that, as observed during one of the nine years studied.

Water-content monitoring at the Beatty site from 1978 to 1980 measured infiltration and redistribution of water down to approximately 2-m depth in 1979 to 1980 (Nichols, 1987). The deep penetration of water was attributed to high precipitation in 1978 (23.35 cm at the site, twice the long-term average) and precipitation in January 1979 (4.62 cm) followed by additional rainfall in March 1979 (2.52 cm). This emphasizes the importance of antecedent water content in the soil and the sequence of precipitation events in controlling percolation. Monitoring at Beatty from 1984 to 1988 showed water movement was restricted to the upper one meter during this period (Fischer, 1992).

Water content was monitored in a small scale ephemeral stream setting (maximum topographic relief of 0.65 m) in the Hueco Bolson of Texas (Scanlon, 1994a). Approximately monthly monitoring of water content from July 1988 to October 1990 showed that within the detection limit of the neutron probe ($\pm$ 0.01 m^3/m^3) water content remained constant from 0.3 to 41 m depth. Although precipitation during 1989 was 50 percent of the long term mean annual precipitation, precipitation during 1990 was similar to the long term mean. Data from these desert basins indicate that penetration of water is often restricted to the shallow subsurface. The reason for the shallow penetration of water in desert soils is the large storage capacity of the surficial sediments. For example, soils similar to those at Ward Valley with an initial average volumetric soil-water content of 10 percent and a water content at saturation of 35 percent have a maximum additional storage capacity of 25 percent by volume, which is equivalent to 25 cm of water storage per meter of soil profile. To bring this into perspective, if an annual rainfall of 12.7 cm could fall in one day, the 12.7 cm of water could hypothetically be stored in only 50.8 cm of soil. In reality, the soil would not come to full saturation, but water would move deeper into the soil profile. Even if the soil-water content were to increase to 17.5 percent (which is 50 percent of saturation), the 12.7 cm annual precipitation would wet the soil down to only 170 cm. This demonstrates the great storage capacity of dry desert soils in general and of Ward Valley in particular.

Potential Energy of Soil Water

In contrast to water content, water potential energy is continuous across different soil types and is typically used to infer the flow direction. Water flows from regions of higher total potential energy to regions of lower total potential energy. In areas with moderate to high subsurface water flux, gravity and matric potential are the dominant driving forces. Matric forces result from the interactions of the water and the soil matrix and include capillary and adsorptive forces. An example of such behavior is demonstrated when a sponge is placed in water. Water can move upward, against gravity, into the sponge until some equilibrium is reached.

Matric potential is expressed in meters of water, bars or megapascals (MPa). For comparison, 1 Mpa = 10 bars = 102 m of water. Because water is tightly held by unsaturated soil, the matric potential has a negative value. If soil becomes wetter, its matric potential becomes less negative until at saturation it becomes zero. Matric potential is related to soil-water content through the soil water retention curve.

Hydraulic Conductivity

The rate of water flow q through soil is proportional to its hydraulic gradient dH/dz with the proportionality constant being the hydraulic conductivity, K. This is expressed through Darcy's law:

$$q = -K \, dH/dz \qquad \text{(Equation 3.1)}$$

where H is the hydraulic head, equal to the sum of the matric potential head and the gravitational potential head, and z is the vertical space coordinate taken as positive upward (see Box 3.1).

BOX 3.1

DRIVING FORCES FOR FLOW IN THE UNSATURATED ZONE

The soil-water potential energy is typically used to infer the water flow direction because water will flow from regions of high to low potential. In the unsaturated zone, many gradients may be important, as is indicated by the generalized flux law (modified from de Marsily [1986]):

$$q = - L_1 \Delta H - L_2 \Delta T - L_3 \Delta C$$

where q is the flux, L_1, L_2, and L_3 are proportionality constants, Δ is the gradient operator, H is the hydraulic head (sum of matric and gravitational potential heads), T is the temperature and C is the solute concentration. This equation holds for systems in which water flow occurs in liquid and vapor phases. The direction of liquid flux is controlled by gradients in hydraulic head whereas the direction of isothermal and thermal vapor flux is controlled by hydraulic head and temperature gradients, respectively. In some flow systems, temperature and osmotic potential gradients are negligible and the flux can be simplified to the Buckingham-Darcy law (the first term to the right of the equal sign in the above equation, the means of calculating downward liquid flow).

To quantify the water flux, information on the proportionality constants is also required. Under isothermal conditions, which implies absence of a thermal gradient, and when liquid flow dominates, the relationship between hydraulic conductivity and water content or matric potential is used to calculate the flux. Hydraulic conductivity decreases exponentially with decreased water content or matric potential. In arid systems, where the matric potential gradient is large, the reduction in hydraulic conductivity with decreased matric potential will outweigh the effect of the increased gradient and will result in negligible flow. In cases where vapor flux is important, isothermal and thermal vapor diffusivities should be measured or estimated to quantify vapor flux.

The ability of unsaturated soil to conduct water, i.e. the hydraulic conductivity, is directly related to its water content. At saturation, soils have their highest hydraulic conductivity. As a soil dries, the larger pores empty first, causing the hydraulic conductivity to decrease. In fact, the hydraulic conductivity of many soils decreases exponentially with decreasing water content or soil-water potential (Gardner, 1958). Thus a decrease in water content by only a few percent can produce a 10-fold or a 100-fold decrease in hydraulic conductivity. Therefore, the very low water contents found in the subsoils of vegetated areas at interstream positions of warm deserts suggest very low hydraulic conductivities and low recharge rates.

In typical interstream settings in arid regions where the soils are extremely dry and water fluxes are low, much of the water movement may occur in the vapor phase. Because the fluxes are so small, the direction and magnitude of the fluxes may be difficult to determine. In addition to liquid flow, as a result of matric and gravitational potential gradients, vapor flow induced by temperature gradients may also be important.

Potential Energy of Soil Water

Darcy's law also requires a knowledge of the gradient in potential energy, dH/dz. Instruments used to measure potential gradients include tensiometers and heat dissipation probes (HDP's), which measure matric potential, and thermocouple psychrometers, which measure temperature and water potential (sum of matric and osmotic potential). Osmotic potentials can be calculated from soil-water chloride concentration data (Campbell, 1985) and can be subtracted from water potential to estimate matric potential.

Except in the shallow subsurface after rainfall, levels of water potential measured in arid and semiarid regions generally decrease toward the surface. This suggests an upward driving force for liquid and isothermal vapor flux (Fischer, 1992; Estrella et al., 1993; Scanlon, 1994). The osmotic component of water potentials, which refers to the energy required to remove dissolved salts from the water, is generally low at depth at these sites because chloride concentrations below the top 10 m are generally low (Phillips, 1994).

Long-term monitoring records of soil-water potentials and water contents for arid sites are limited. Monitoring data from the Beatty site displayed increased water potentials between depths of 1 to 2 m after precipitation in November 1987 (Nichols, 1987). Neutron moisture probe readings for the same period of record did not show changes in water content, probably because the changes in water content were within the standard error of the neutron probe readings. In the Chihuahuan Desert of Texas, at a silty loam site within a small-scale ephemeral stream setting, water potentials measured from summer 1989 to summer 1990 were out of range (<-8.0 MPa) of the thermocouple psychrometers in the upper 0.8 m, because the sediments were too dry (Scanlon, 1994). Water potential monitoring has continued since that time to 1995 and has shown that the wetting front has not penetrated below the upper 0.3 m at this site. In general, below the shallow zone of active circulation, matric potentials show seasonal fluctuations in response to seasonal temperature fluctuations down to depths of 12 m (Fischer, 1992; Scanlon, 1994). Below the zone of seasonal fluctuations, water potentials remain constant over time.

One can also estimate the matric potential that would exist in the unsaturated zone if it were in hydraulic equilibrium with the water table. Under equilibrium conditions, the total soil-water potential energy is everywhere constant. (In the sponge example discussed previously, the capillary forces drawing the water upward are in balance with the force of gravity tending to pull the water downward). Under these conditions, the soil matric potential energy is at equilibrium with gravitational forces. If z (the vertical space coordinate) is taken as positive upward and zero at the water table, then under these conditions the equilibrium soil matric potential can be described by a line starting at zero at the water table and equal to the negative of the height above the water table (Figure 3.2). Under steady flow conditions, matric potentials displaced to the right of the equilibrium line indicate downward flow, whereas those plotted to the left indicate upward flow (Figure 3.2). At several arid western sites (Fischer, 1992; Estrella et al., 1993; Scanlon, 1994) matric potentials plot to the left of the equilibrium line indicating upward driving forces for liquid and isothermal vapor movement (Estrella et al., 1993; Scanlon, 1994). At the Nevada Test Site (NTS) this upward driving force is restricted to the upper 40 m. Below this depth, water potentials plot to the right of the equilibrium line, which suggests that water in this portion of the profile may be moving downward to the water table (Sully et al., 1994).

Below the zone of seasonal temperature fluctuations, the upward geothermal gradient provides an additional upward driving force for thermal vapor movement. This is approximately 0.06°C/m at the Beatty site (Prudic, 1994a). Sully et al. (1994) suggest that this upward thermal vapor flux may be of the same order of magnitude as the downward liquid flux at depth at the NTS; such a condition would result in no net flux of water but a slow downward migration of water-soluble chemicals.

Vegetation also plays a critical role in removing water from desert soils. Lysimeter data from Hanford and Las Cruces showed that deep percolation from bare sandy soils can range from 10 to >50 percent of the annual precipitation (Gee et al., 1994). Therefore, the use of data developed on undisturbed sites in Ward Valley to predict long-term performance through the operation and closure period is acceptable only if the conditions include reestablishment of the active plant communities after the waste is emplaced. There remains uncertainty of the effects of trench construction on the long-term water balance at any disposal site, and the goal of the closure and trench cover program outlined in the license application is to reduce these uncertainties.

Nature of Water Movement: Piston Flow and Preferential Flow

In general, two types of water movement occur in unsaturated zones, piston flow and preferential flow. Piston flow refers to uniform water movement downward through the soil matrix. Infiltrated water displaces initial water. In contrast, preferential flow refers to nonuniform downward water movement along preferred pathways, such as continuous vertical fractures, or pathways created by changes in soil characteristics.

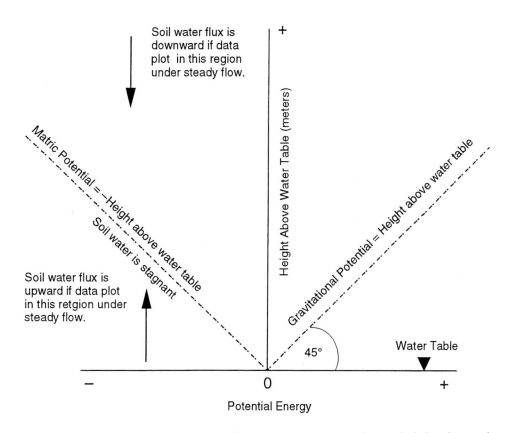

Figure 3.2 The relationship of the equilibrium matric potential to height above the water table. If the soil water is stagnant, the matric potential is exactly balanced by the gravitational potential represented by the height above the water table. A simple measure of the direction of flow can be made by plotting matric potential head (calculated from field measured water potentials by subtracting the osmotic potential) versus the height above the water table.

A variety of factors are important in preferential flow, including continuity of preferred pathways and exposure of the pathways to ponded or perched water (Beven, 1991). Because water will enter fractures (even microfissures and relatively small vertical cracks) only when soils approach saturation, preferential flow has been documented mostly in more humid regions with higher rainfall. The Ward Valley alluvium consists of variable mixtures of poorly sorted sand, silt, gravel, and clay, with greater variability vertically than horizontally, as discussed later in this chapter. Thus the most significant pathways are likely to be along root tubules, cracks, and fracture planes (including faults). Although some have considered flow associated with major hydrologic features such as ephemeral streams and playas as preferential flow (Gee and Hillel, 1988), we will restrict our discussion to the kinds of pathways described above.

In uniform soils and at low rates of soil-water movement, water will generally flow in a piston-like manner, particularly if the soils are dry initially. Under piston-like flow conditions most, if not all, preexisting water ("old" water) is displaced and moved ahead of the "new" infiltration water added from above. Under certain conditions, only a fraction of the "old" water is displaced and water moves through preferential pathways. Preferential flow is the process whereby water and solutes move along preferential pathways through a porous medium (Helling and Gish, 1991). During preferential flow, local wetting fronts may propagate to considerable depths in a soil profile, essentially by passing the matrix pore space. (Bevin, 1991; Steenhuis et al., 1994). Preferential flow has been demonstrated under laboratory conditions (Glass et al., 1991) and under simulated rainfall conditions (28-152 cm/day) in forested field soils (Turton et al., 1995). However, in dry desert soils, water that is moving through preferential pathways, such as roots and root tubules, is expected to be absorbed by the dry soil around the pathways. Also, layering caused by differences in texture will disrupt the continuity in the flow paths. Therefore preferential flow is expected to be damped out over a relatively short distance below the root zone.

The rate and depth of infiltration along preferential pathways is often greater below areas of surface ponding or frequent flooding than below higher ground.

Piston flow is generally found in arid sites where sediments are not subjected to intermittent or continuous ponding. At study sites in both southern New Mexico and southern Nevada, where long-term monitoring of soil-water content has been conducted, measurements have confirmed that significant changes in water content are restricted to the upper 1-2 m. The water input to these systems that receive water only from precipitation may be too low to result in preferential flow. In addition, much of the water in these dry soils is adsorbed on the grain surfaces and therefore cannot move along preferred pathways.

Piston flow is also indicated by single peaks on profiles of the chemical tracer, chlorine-36 (^{36}Cl), as in the profile measured in a small-scale ephemeral stream setting in the Hueco Bolson in Texas (Scanlon, 1992a). The single ^{36}Cl peak suggests that the bomb related ^{36}Cl which fell on the land surface in the mid 1960's has moved downward uniformly and retains its bell-shaped profile. If non-piston flow had occurred, the ^{36}Cl in the soil zone would show a much more erratic pattern, with alternating highs and lows corresponding to zones of fast and slow water movement. Large-scale infiltration experiments conducted at Las Cruces showed that the movement of dissolved solids, or solutes, lagged behind the water, or wetting front (Young et al., 1992). The lag between the solutes and wetting fronts increased with depth, which indicated piston displacement of initial soil water, rather than preferential flow. The lag also increased with increased initial water content.

Previous studies specifically identifying preferential flow in alluvial deposits in desert regions are limited. Allison and Hughes (1983), in a study of recharge in south central Australia, observed tritium much deeper (12 m) in the soil profile than was predicted from other tracers and the general aridity of the region. They attributed the tritium transport to flow along channels occupied by living roots. No data on tritium transport below the root zone were presented (Allison and Hughes, 1983). Shrubs at Ward Valley, such as creosote bush have much shallower rooting depths ($\leq$ 1-2 m) than the eucalyptus plants reported by Allison and Hughes (1983). Fracture flow has also been reported at Yucca Mountain, Nevada by Yang (1992) and Fabryka-Martin et al. (1993); however, the geologic setting at Yucca

Mountain differs greatly from that at Ward Valley, as the Yucca Mountain unsaturated zone is in faulted and fractured volcanic rock.

Preferential flow has also been found in fissured sediments in the Chihuahuan Desert of Texas (Baumgardner and Scanlon, 1992; Scanlon, 1992a). They conducted detailed studies of fissured sediments in the Hueco Bolson in Texas. These features consist of an alignment of discontinuous surface gullies (≤140 m long) underlain by sediment-filled fractures (2 to 6 cm wide) that extend to a depth of ≥6 m. Similar features have not been found at Ward Valley.

Fractures were observed in the Quaternary alluvium at the site and are prominently exposed in the old I-40 borrow pit. These could be either desiccation or tectonic fractures. Their nature and continuity at depth are unclear, but unless a systematic investigation is made of these fractures, it is not possible to assess their origin or continuity. The impact of fractures on fluid flow can be deduced from the distribution of water content and tracers found in the unsaturated zone.

In summary, studies have shown that water movement in arid soils is controlled by several factors, such as energy gradients, topography, and the complexity of the soil materials. It is therefore critical that comprehensive and detailed studies be performed at any site where the unsaturated zone is to be used as a primary barrier for waste isolation. In the next sections, the data for the Ward Valley site are reviewed with respect to this criterion.

THE NATURE, DIRECTION, AND MAGNITUDE OF WATER FLUX BENEATH THE WARD VALLEY SITE

The Wilshire group (Wilshire, 1993a,b, and 1994) questioned the assertion in the license application that percolation is negligible beneath the site. They based this on observations of preferential flow in other areas and on the reported tritium concentrations found at depths to 30 m below the surface. They also raised the question of the potential for recharge in Homer Wash and its implication for site performance.

The techniques commonly used to estimate soil-water flux in the unsaturated zone fall into two broad categories: soil physics methods and chemical tracer techniques.

Soil Physics Methods Applied to Ward Valley

The two types of soil physics methods used to compute deep percolation at the Ward Valley site are: water-content monitoring and hydraulic gradients.

Water-Content Data

Water contents measured from the surface to 27 m depth in 82 soil samples collected from six borings (GB-1 through GB-6, LA Appendix 2420.B) were generally very low. Twenty samples had volumetric water contents less than 5 percent, 57 samples had water contents between 5 and 10 percent, and 5 samples had water contents between 10 and 15 percent. The higher water contents were in the clayey sands. These data suggest that the sediments are extremely dry.

Soil-water content monitoring at the Ward Valley site was very limited. Neutron-probe logging of the water content was conducted under ambient conditions in only one access tube to a depth of 6 m. A pneumatic drilling system (ODEX) was used to install the access tube and consisted of drilling a 15-cm diameter borehole and installing a 5-cm diameter aluminum access tube. Native sediments were used to backfill the annular space surrounding the access tube. A baseline log obtained on October 5, 1990 serves as a standard to compare to later logging. One standard deviation of the count rate difference between the baseline and observed neutron logs was approximately 95 counts per second, which corresponds to an uncertainty in the volumetric moisture content of ±4.5 percent (U.S. Ecology, 1990). This value of 4.5 percent is large relative to typical standard errors reported (± 1 percent) for neutron probe logs.

Water-content monitoring using the neutron probe was conducted for a 12-month period that ended on August 27, 1990. Three significant rainfall events occurred during this period; January 2 (0.71 cm), May 28 (2.49 cm), and July 13 through 15 (5.80 cm), 1990. The soil was excavated to allow visual observation of the depth of the wetting front after these events and showed that water infiltrated to 30 cm on January 5, 51 cm on June 1, and 90 cm on July 18. An increase in water content up to 19.8 percent was monitored after the July rain at one meter below the land surface. Monitoring data did not indicate water movement below this depth. The shallow penetration of the wetting front is attributed by USE to the large storage capacity of the surficial sediments.

Conclusions Regarding Water-Content Data from the Ward Valley Site

Water contents from 82 soil samples collected near the surface to a depth of 27 m from 6 boreholes were low (94 percent of the samples had water contents less than 10 percent, and 6 percent of the samples had water contents between 10 and 15 percent) which suggests that the sediments are extremely dry and that subsurface water fluxes are very low. The absence of water-content fluctuations below the root zone is consistent with the license application conclusion that deep percolation is extremely small. However, the short monitoring period (one year) and monitoring of only one neutron-probe access tube does not provide reasonable assurance of very low infiltration over the entire site. Further lines of evidence supporting this conclusion are necessary.

Monitoring of Water Potential at Ward Valley

Thermocouple psychrometers were installed at approximately 3 m intervals between 2.4 m and 30 m depth in April 1989 to monitor water potentials and temperature in three adjacent boreholes. Four thermocouple psychrometers were installed at each depth. Considerable differences were often observed among these quadruple psychrometers. Reported water potentials ranged from approximately -3 to -6 megapascals (MPa), a measure of pressure, which are consistent with the low water contents measured. Although the backfill sediments were dry, the small borehole diameter (15 cm) should have minimized the equilibration time of the backfill sediments with the native sediments.

The one-year monitoring period began in August 1989. Fluctuations in ambient air temperatures affected measurements of both subsurface temperature and matric potential. Attempts made in the field to minimize the effect of variations in ambient air temperature on the data logger were not successful. Regression analysis to correct for the effect of ambient air temperature variations reduced the variation in temperature and water potential with time but could not completely remove the problem. Therefore, vertical temperature gradients cannot be estimated from the data. Temperature variations also affected water potential measurements. Use of the corrected temperatures based on regression analysis reduced the seasonal water potential fluctuations but did not eliminate them. The temperature problem may result in part from the use of a low-sensitivity data logger (HP-115 data logger, 100 to 150 nV resolution) that was not adequately designed to minimize thermal gradients (R. Briscoe, 1994). It is difficult to evaluate equilibration of the psychrometers because of the problems with ambient air temperature effects, and it is inappropriate to use these data to calculate a gradient within the upper 30 m.

Direction of Water Flow

Rather than a direct calculation of the gradient between water potentials measured by the psychrometers at different depths, an estimate of the direction of water flow can be made by comparing the range of calculated matric potentials in the upper 30 m to that which would be expected if no liquid water were moving. As discussed earlier, under the conditions of no water movement, the matric potential at any point in the soil would be equivalent to the height above the water table (when expressed in the correct units when z is taken as positive upward and zero at the water table). If no flow were occurring at the Ward Valley site, we would expect matric potentials to range from -1.7 to -2.0 MPa in the upper 30 m given the depth to water as determined from the adjacent monitoring wells. We must first remove the osmotic potential from the measurements of water potential. Osmotic potentials calculated from soil-water chloride concentrations in boreholes GB-1, 2, and 4 (Prudic, 1994b), according to procedures outlined in Campbell (1985), ranged from 0 to -2 MPa and were lowest in near-surface sediments and ≤ -1 MPa below 5-m depth. Subtraction of osmotic potentials from measured water potentials resulted in matric potentials of -3 to -4 MPa, that plot to the left of the equilibrium matric potentials (Figure 3.2) and suggest a net upward water flux under steady flow conditions. **The water-potential gradient cannot be determined from the Ward Valley data because of the absence of accurate site temperature data and lack of correspondence between water potentials recorded by quadruple psychrometers installed at the same depth. However, water-potential gradients in interstream settings at many other arid sites are upward in the upper 40 m (Fischer, 1992; Estrella et al., 1993, Scanlon, 1994); this evidence suggests that gradients in the upper 30 m at the Ward Valley site are likely to be upward also.**

Because of the dry condition of the sediments at Ward Valley, as indicated by low measured water contents and water potentials, much of the water movement may occur in the vapor phase rather than in the liquid phase. If vapor-phase movement dominates, temperature gradients may be important in controlling the direction and rate of water movement. Temperatures monitored at Ward Valley were affected by ambient air temperature fluctuations. Data from Beatty, Nevada (Prudic, 1994a) and Hueco Bolson, Texas (Scanlon, 1994) indicate that seasonal temperature fluctuations extend to a depth of approximately 12 m. Numerical simulations of non-isothermal flow at Hueco Bolson indicate that in the zone of seasonal temperature fluctuations (in the upper 10 m), higher soil temperatures near the surface in the summer should cause a net downward thermal vapor flux on an annual basis because of the higher thermal vapor diffusivities at the higher temperatures in the summer (Scanlon and Milly, 1994). Below the zone of seasonal temperature fluctuations, upward geothermal gradients result in upward thermal vapor flux. The upward geothermal gradient measured at Beatty is 0.06 °C/m (Prudic, 1994a).

Heat Dissipation Probe Measurements of Matric Potential

Heat dissipation probes (HDPs) were also used to monitor matric potentials in the soil profile. They were installed in wet silica flour at depths of 0.6 m, 1.2 m, 2.4 m, and 4.8 m and showed matric potentials ranging from approximately -0.2 to -0.5 MPa. The HDPs were not very reliable, as only 4 of the 8 HDPs operated throughout the monitoring period.

Estimated volumetric water content from these matric potentials based on water-retention data from 10 samples is 15.5 to 16.9 percent, which is much higher than the laboratory-measured values for these sediments (3 to 8 percent) (U.S. Ecology, 1990). The much higher matric potentials measured by the heat-dissipation probes relative to water potentials measured by the thermocouple psychrometers are attributed to wet silica flour used in installation of the HDP's. Use of wet silica flour results in a much longer equilibration time needed for the instruments to measure the actual soil conditions. The HDPs, however, did record infiltration as indicated in the matric potential increase from -0.4 to -0.05 MPa at 0.6 m after the July 1990 rainfall event.

Conclusions Regarding Water Content and Water-Potential Data at the Ward Valley Site

The license application concludes that the gradient for liquid movement indicates that water is currently moving upward. This conclusion is based on field data collected in the upper 30 m of the unsaturated zone. No water potential or gradient data were collected in the unsaturated zone below 30 m.

The committee concludes that the water contents and water potentials are low, which indicates that the sediments are dry and that the subsurface water fluxes in the upper 30 m are extremely low. The license application states that the water-potential gradient is upward, but, in the committee's view, the Ward Valley data are not of sufficient quality to justify this conclusion. However, the fact that the water potentials are lower (drier) than those that would be predicted, based on equilibrium or no flow conditions, suggests that flow is upward in the upper 30 m under steady flow conditions. The direction of flow below 30 m cannot be determined from the license application because no water potential data are available below that depth. However, because of the low water contents measured at 30 m, water fluxes are expected to be extremely low below 30 m at the site.

Although there were several problems with instrumentation for water-potential monitoring, all recorded water-potential values were very low (-3 to -6 MPa). These water-potential values are consistent with the measured water content values when examined using water retention functions measured for site soils. Matric potential measured by HDP's were much wetter and are not consistent with measured water contents. The higher matric potentials measured by the HDP's, however, are attributed to the use of wet silica flour during HDP installation, which resulted in very long equilibration times.

ENVIRONMENTAL TRACERS AS INDICATORS OF SOIL-WATER MOVEMENT

Environmental or chemical tracers are commonly used in soil and ground-water research to investigate the direction of soil-water movement. Tracers represent a generally spatially uniform (to first order) input to the soil and ground-water system. In many cases, the history of the tracer input is well known or at least can be constrained within reasonable bounds (Allison et al, 1994). The two tracers commonly used, chloride and tritium, also have the advantage that they can be readily analyzed.

Tracers are not, however, perfect indicators of water movement. The tracers offer only an indirect indication of water movement. Processes that affect infiltration can have significant influence on the distribution of tracers. The tracer may not always move exactly as does the water, so multiple tracers are often needed to investigate the unsaturated zone.

Chloride as a Soil-Water Tracer

Chloride has begun to gain wide acceptance as an indicator of recharge and water movement in soils (Allison and Hughes, 1983). Soluble chloride, present in both precipitation and in dust (or dryfall), enters the soil at the land surface. Other sources of chloride in the soil, such as chemical weathering, are generally minor. Water passing across the soil/atmosphere interface carries the soluble chloride into the soil profile. Plant roots extract some portion of the infiltrated water, selectively leaving behind chloride ions in the soil-water solution. Evaporation at the land surface can also concentrate some salts at the surface in the form of efflorescent crusts, although this is less common in vegetated areas. These two processes, water carrying the chloride down and evapotranspiration, lead to a progressive enrichment of chloride in the root zone.

The time required to accumulate large amounts of chloride in desert soils is extremely long (perhaps thousands of years); however, most of the chloride can be flushed out of the soil if the water moves rapidly through the profile once as a result of ponding or increased infiltration. The absence of chloride in soil water indicates either that water fluxes are sufficiently high to minimize chloride accumulation or that high water fluxes flushed out accumulated chloride. Because chloride is readily flushed out of the soil if subsurface water fluxes are high, the occurrence of high chloride concentrations is very good evidence of low water fluxes for very long time periods. Chloride profiles measured in small-scale ephemeral stream settings in the Hueco Bolson in Texas are characterized by high maximum chloride concentrations (up to $9,000 \text{ g/m}^3$ (9 gm/l)) which indicate that subsurface water fluxes in these small-scale stream settings are negligible (Scanlon, 1991). The maximum bomb pulse chlorine-36 concentration in the same setting was measured at 0.5 m depth, which also indicates very low water fluxes and corroborates the meteoric chloride data (Scanlon, 1992a).

The two processes that move chloride downward below the root zone are diffusion and advection. Chloride (and any other salts excluded by the plants) will diffuse downward to the water table as a result of concentration gradients. Figure 3.3a shows a typical profile of chloride undergoing diffusional transport. The highest concentrations are directly beneath the root zone, with concentrations tapering to the water table. The root zone remains low in chloride, presumably due to periodic infiltration of precipitation which flushes any salts that have diffused upward.

Advection, or percolation, of water below the root zone is the amount of water that escapes evapotranspiration and descends to the water table. It carries with it the dissolved Cl⁻ and other ions and salts. With no further mechanism for enrichment below the root zone, the concentration of chloride remains uniform down to the underlying ground water, where, if downward percolation is the only source of recharge to the aquifer, the soil water and ground water will have the same concentration. If the ground water is recharged primarily from elsewhere, the soil water and ground-water concentrations do not necessarily coincide. Figure 3.3b shows a typical chloride profile undergoing advective transport.

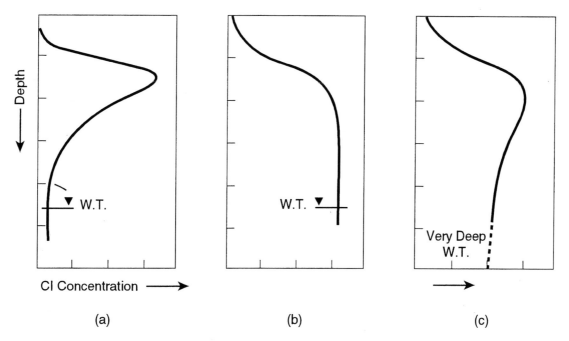

Depth

Cl Concentration ⟶

(a)   (b)   (c)

W.T.   W.T.   Very Deep W.T.

Figure 3.3 Schematic depth profiles of chloride concentration in soil water: (a) Extraction of water by plant roots followed by either diffusive loss of chloride to the water table or preferred flow of water through and below the root zone; (b) Piston flow of water with extraction of water through the root zone; (c) A chloride profile reflecting past recharge conditions; W.T. = water table (after Allison et al., 1994).

Under these conditions, the rate of recharge or percolation, R , can be calculated as:

$$R = J_{Cl}/C_{SW}$$ (Equation 3.2)

where J_{Cl} is the chloride flux at the land surface and C_{SW} is the concentration of chloride below the root zone.

This technique, formally known as the Chloride Mass Balance (CMB) (Allison and Hughes, 1983), is now widely accepted in soil-water studies to estimate recharge in arid climates. As seen in equation 3.2, two parameters are needed, the chloride flux to the land surface and the concentration deep in the profile. While the latter is relatively straightforward, estimating or measuring the chloride flux is more difficult and is the source of greatest uncertainty. In studies of salt chemistry in rainfall, Junge and Werby (1957) show that chloride concentrations are inversely proportional to the distance from the ocean, the primary source of salts. Dryfall or dust fallout can also be a major contributor to the total flux of chloride to the land surface. In a study of ground-water recharge in the southern Great Basin,

Dettenger (1989) reported up to 50 percent of the total salt flux was due to dry deposition. Thus, it is critical to account for all sources of chloride for an accurate measure of recharge

A key assumption in using the CMB to estimate downward flux is that the entire soil profile is at equilibrium, i.e. the influx of chloride to the soil surface is exactly balanced by the discharge of chloride to the water table. In shallow soils or in agricultural settings, such an assumption is often justified. In deep arid zone profiles, however, the changes in climate, vegetation, and recharge conditions, combined with the slow rates of recharge, often lead to profiles that are not in equilibrium. In such cases, the distribution of chloride in the deep, unsaturated zone cannot be used directly to infer recharge rates at the present time.

The transient recharge conditions previously described would not be an unreasonable model for the Ward Valley site. The region has undergone dramatic climatic fluctuations (Hostettler, 1994) with periods of higher precipitation, and it is unlikely that the chloride profile is at steady state. Under such conditions, the ideal chloride profile might display a pattern of chloride concentration similar to that shown in Figure 3.3c. Within the root zone, the chloride concentrations would be low, reflecting repeated flushing by precipitation. Below the root zone, the concentration would quickly reach a maximum, reflecting the current aridity. Below the maximum concentration, the profile would decrease as is typically seen in many southwestern United States soils (Phillips, 1994). Profiles of this shape have been interpreted by Scanlon (1992) and Phillips (1994) to be the result of the end of Wisconsin glaciation (last of the Pleistocene glacial stages, about 11,000 years ago), when the climate was cooler and wetter.

In the case of the non-steady state profile, the application of the CMB is not appropriate. Rather, the accumulation of chloride can be used to estimate the time required to develop the profile. If the chloride is acting as an ideal tracer for the water, the age of the water at any depth can be estimated. Several workers have applied this technique to profiles in the southwest to infer recharge conditions over the last 10,000 to 20,000 years (Cook et al, 1992; Scanlon, 1992; Phillips, 1994; Prudic, 1994b). The time, t, needed to accumulate the chloride to any depth, Z, is given by:

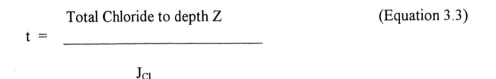

$$t = \frac{\text{Total Chloride to depth Z}}{J_{Cl}} \qquad \text{(Equation 3.3)}$$

where the total chloride is calculated from the chloride concentration in the soil multiplied by the water content volume, and J_{Cl} is chloride input per unit time.

The chloride accumulation technique provides an integrated measure of the recharge and water flux through the unsaturated zone. While the technique is robust, it must be viewed in light of the assumptions used in its development. Clearly, the chloride accumulation rate must be well known or constrained for long periods of time. The chloride must be an ideal tracer of liquid water, i.e., vertical piston flow is required in the soil. Each of these assumptions is discussed below to assess their impacts on the outcome.

Chloride Accumulation Rate

The chloride accumulation rate is the rate at which chloride falls on the land surface in precipitation and dust. If the chloride concentration in the precipitation is (to first order) controlled by the distance to the ocean, its concentration at the Ward Valley site is unlikely to vary significantly with time. Under the current climatic regime, the precipitation at the site is approximately 12 cm/yr. This may have been up to 40 percent higher during the last 20,000 years (Hostettler, 1994). Under this scenario, the chloride flux due to precipitation could, conservatively, have been as much as 40 percent greater than the current rate. With this, we introduce an uncertainty of a factor of two in the age estimates.

Dry deposition of dust and salts will surely have varied over both short and long time scales. Studies in Nevada (Dettenger, 1989) suggest that the current chloride flux from dryfall is of the same magnitude as that brought in by precipitation. This can be taken only as a general rule as some areas could receive much higher or lower fluxes depending on local settings, elevation and proximity to local sources of chlorides. With the desiccation of many Mojave lakes at the end of the Pleistocene, it is likely that the dryfall component was greater than today. On the other hand, the presence of ^{36}Cl in deep profiles at or near modern ratios found by Phillips et al. (1988) and Scanlon (1992) in other southwest U.S. sites would suggest that chloride dryfall did not dramatically increase. The dryfall question remains open and must be answered before confidence can be given to precise dating with chloride. The use of chlorine isotopic studies, both stable and cosmogenic, in the soil water may be very useful in this regard, as many saline lakes have ^{36}Cl signatures very different from that of modern precipitation.

Non-Piston Water Movement

The chloride age requires piston flow for it to represent the age of the soil water, although, as described previously, field studies in humid climates have shown that water can migrate along preferential pathways created by roots, fractures or variations in soil texture. In such cases, some portion of the infiltrating water moves quickly through the soil, bypassing much of the soil matrix in the upper soil profile. Such events can lead to a chloride profile that also shows a chloride bulge near the surface. In this case, changes in the recharge rate did not produce the distribution, but rather the occasional introduction of low salinity water bypassing much of the upper soil matrix deeper in the profile produced the observed pattern. It is, therefore, not possible to assess the nature of soil-water movement on the basis of the distribution of chloride alone. Studies of chloride profiles must include an analysis of the likelihood for non-piston movement using other tracers or hydraulic indicators.

Chloride Data from Ward Valley

Chloride data from the Ward Valley site are discussed by Prudic (1994b). Chloride concentrations were measured in soil samples collected from six geotechnical borings (GB-1 to GB-6) to a depth of 30 m (LA Appendices 2500.A and 2600.A) (See Figures 2.4 and 2.5 for borehole locations.) No chloride data are available below a depth of 30 m. Bulk density and volumetric water content were not measured in the samples that were analyzed for chloride, so chloride concentrations in soil water were calculated using bulk density and volumetric water content from adjacent samples by Prudic (1994b) (Figure 3.4)

The chloride concentrations in the soil water were very high by comparison with localities with less stable surfaces or high precipitation, which suggests that the net subsurface water flux is small. If downward fluxes had been high, chloride would have been flushed out of the soil. Three of the boreholes yielded insufficient data to evaluate the depth distribution of chloride (one sample from GB-3; 4 samples from GB-5 and 3 samples from GB-6). Therefore, data from these boreholes will not be included in the following discussion. The chloride profiles are generally bulge-shaped with low chloride concentrations

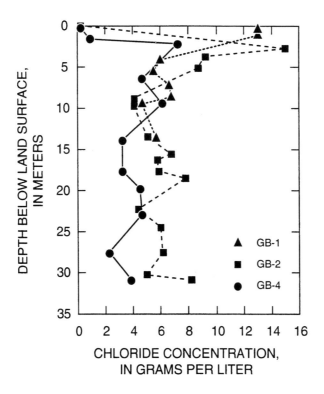

Figure 3.4   Chloride concentrations in pore water of unsaturated sediments from three boreholes at the Ward Valley site. Borehole locations shown on Figures 2.3 and 2.4 (Prudic, 1994b).

in soil water in the surficial sediments and maximum concentrations measured at 2.4 to 2.9-m depth in GB-2 (15 grams per liter (g/l)) and at 2.1 to 2.3 m in GB-4 (7.3 g/l) (Figure 3.4). Data from GB-1 differ from other data in that the maximum concentration in soil water was measured in the surface sample in GB-1 (at 0 to 0.3 m depth; 14 g/l).

Chloride concentrations in soil water decrease below the maximum to levels that range from 3.8 g/l in GB-4 to 8.2 g/l in GB-2. Prudic (1994b) calculated chloride ages based on data from boreholes GB-1 and GB-2 using a chloride deposition rate of $1.64 \times 10^{-5} g/cm^2/yr$ (Prudic, 1994b). These ages increased from about 10 yr at 0.3 m to 52,000 to 58,000 yr at 30 m depth. Water fluxes estimated from the chloride data ranged from 0.03 mm/yr for GB-1 and GB-2 to 0.05 mm/yr for GB-4 below 10-m depth.

The two possible explanations for the bulge-shaped chloride profiles in Ward Valley are: (1) paleoclimatic variations in recharge with higher recharge in the Pleistocene and little or no recharge in the Holocene or (2) small-scale preferential flow diluting chloride below the peak. **In the view of the committee, small-scale preferential flow is unlikely because of the dry nature of the sediments, with most water tightly adsorbed onto the grain surfaces. While preferential flow cannot be entirely ruled out, the high chloride concentrations suggest that percolation is extremely low at the site.**

The likelihood of relatively high recharge during the Pleistocene with little or none in the Holocene is not incompatible with ages of ground water below Ward Valley that are considerably younger than the "chloride accumulation age", because the saturated-zone ground-water recharge zone was probably not in the immediate vicinity of the site.

Conclusions Regarding Chloride Data from the Ward Valley Site

The committee regards the chloride data analyzed from the three boreholes as evidence that percolation at the site is very small through the unsaturated zone. Percolation probably was very small in the late Pleistocene (30-10 ka) as well, based on the high chloride concentrations below the root zone evaluated by Prudic (1994b). The committee regards the bulge profile shape as most likely the result of changes in the climatic and recharge regime between 10 and 20 ka.

Tritium as a Soil-Water Tracer

Tritium, T or ^3H, a radioactive isotope of hydrogen, has been used to trace soil- and ground-water movement since the early 1960's (Vogel and Ehhalt, 1963; Clayton and Smith, 1963; Münnich and Roether, 1963; Vogel, Ehhalt, and Roether, 1963). The basis for its use is the thermonuclear bomb testing during the late 1950's and early 1960's which produced relatively high tritium levels compared to the natural cosmogenic levels. The bomb tritium became incorporated into water molecules in the upper atmosphere as HTO, and joined the hydrologic cycle to provide what has become effectively a global radioactive tracer experiment. Since about 1963, however, bomb tritium has been decreasing regularly, and is now close to "natural" cosmogenic levels.

In soil-water studies, researchers have generally relied on the location of the peak concentration to infer downward percolating water. Tritium concentrations are often reported in terms of "Tritium Units", or TU's, where 1 TU = ^{3}H/H x 10^{-18} = 3.2 picocuries/kg water. The half-life of tritium is 12.3 years. Thus, after one half-life, the concentration of tritium in a parcel of water would be half its original value, after two half-lives, 1/4 of the original value, etc. At its maximum in 1963, thermonuclear tritium exceeded 1000 TU in continental precipitation (Fritz and Fontes, 1980). Today's precipitation contains about 5 to 15 TU, depending on location and meteoric conditions.

Limitations of Tritium as a Tracer

Collection of water for tritium analysis from dry soils typical of interfluvial settings in arid and semi-arid regions can be extremely difficult. Analysis of tritium can be done with as little as 4 mL of water (H. G. Östlund, Written Communication, 1989), or as much as 300 mL, depending on the technology used and the precision and detection limit required. This amount can be difficult to collect from dry soils. When water fluxes are very low, much of the water containing recent tritium may still be found in the root zone. In the root zone, velocities can be much higher than in the underlying sediments. As a result, the calculated water velocity may provide an upper limit for the deep soil-water velocity (Tyler and Walker, 1994). If the bomb-produced tritium is solely contained in the root zone, however, this is a strong indication that percolation is minimal. Under conditions of non-piston movement, the tritium from recent infiltration may be spread vertically throughout much of the unsaturated zone and no peak can be identified.

Tritium dating of soil water is further complicated by the fact that HTO can move both as a liquid and as a gas. Even if liquid movement in the unsaturated zone is negligible, gas-phase transport can result in downward movement of tritiated water vapor. The presence of liquid water in the soil significantly reduces this movement (Phillips et al., 1988). As the tritium diffuses downward in the vapor phase, some tritiated water molecules exchange with water molecules in the liquid until they reach equilibrium. Because of the great disparity in the densities of water vapor (approx. 0.02 kg/m^3) and liquid water (approx. 1000 kg/m^3), most of the gas-phase tritium is taken up by the liquid through diffusive exchange. As a result, downward movement of tritium by diffusion in the gas phase should be greatly attenuated in most unsaturated zones.

Tritium Distribution in the Unsaturated Zone

Although the use of tritium in deep, unsaturated zones to infer rates of soil-water movement has several limitations, the distribution of tritium in the unsaturated zone can be used to derive information on the nature of water movement.

In general, tritium is distributed in three distinctive patterns. Under conditions of uniform or piston flow, one would expect to find a tritium peak perhaps slightly displaced from corresponding liquid water molecules infiltrated at the same time. If vapor phase dominates liquid-phase flow, the HTO may precede the H_2O. But if liquid-phase flow dominates, HTO exchange with H_2O that is tightly bound to clays and other silicates will retard the tritium. If the tritium peak were found well within the root zone or just slightly below it, one could conclude that the percolation rate is low. Quantification of exactly how low, however, would require the use of other tracers and methods. If the tritium peak were well below the root zone, a simple estimate of the water velocity could be obtained by taking the distance from the land surface to the peak and dividing it by the number of years that had elapsed since the peak in tritium in the rainfall.

A third possible tritium distribution exists. In studies of percolation in Australia, Allison and Hughes (1983) measured tritium down to 10 m, well below the depth implied by other tracers and the general aridity of the region. They attributed the appearance of such deep tritium to some form of preferential or non-piston flow along root channels. A similar phenomenon has also been reported for fractured rock at Yucca Mountain, Nevada (Fabryka-Martin et al., 1993). These profiles generally do not show a well-defined peak and often exhibit a very erratic distribution; certain depths contain no measurable tritium while tritium is found in high levels immediately below. Such distributions of tritium, therefore, indicate water, in discrete zones, has had little time to be diluted by the surrounding soil water.

Tritium Measurements at Ward Valley

From Table 3.1 it is apparent that the highest tritium values were from water-vapor samples collected from the near-surface air piezometers. Also apparent from this table are finite tritium levels (greater than twice the ± figure) collected at depths as great as 30 m below the ground surface. Tritium detection at depth in the Ward Valley alluvium has been interpreted by the Wilshire group (Wilshire et al., 1993b) to be meteoric water labeled with "bomb" tritium, infiltrated to the collection depths. The license application states that the most probable explanation for tritium at this depth is gaseous diffusion. This committee concludes that neither explanation is correct, and that, as discussed below, the presence of tritium is most likely an artifact of the collection procedure. It is important to the discussion below to point out that the ± figures quoted in Table 3.1 do not indicate the total uncertainty for tritium in soil water, as implied in the footnote. Why this is true requires some background in the concepts of precision and accuracy.

First, accuracy and precision are related but distinctly different quantities. Accuracy of a measurement relates to how well calibrated an instrument is with respect to some absolute or defined standard. Precision is a quantitative statement of the uncertainty of the measurement. Tritium and ^{14}C laboratories usually report only the precision of the laboratory measurement in terms of a number preceded by "±".

Table 3.1   Tritium Results for Unsaturated Zone Soil Vapor[1]

Date	Air Piezometer	Sample Depth (feet)[3]	Tritium Value[2] (tritium units)
6/26/89	GB-1	21.5	1.39 ± 0.57
6/26/89	GB-1	35	1.72 ± 0.51
6/7/89	GB-1	60	0.74 ± 0.41
6/6/89	GB-1	99.5	-0.01 ± 0.58
5/6/89	GB-4	16.5	5.60 ± 0.37
5/7/89	GB-4	58	1.37 ± 0.49
5/7/89	GB-4	99.7	1.02 ± 0.33
6/2/89	GB-4	16.5	6.00 ± 0.72
6/3/89	GB-4	58	1.15 ± 0.66
6/4/89	GB-4	99.7	1.18 ± 0.54
6/24/89	GB-6	18.5	3.94 ± 0.73
6/23/89	GB-6	33	2.07 ± 0.89
6/23/89	GB-6	59	1.38 ± 0.62
6/16/89	GB-6	99.7	1.66 ± 0.39
	Air Moisture Sample		6.91 ± 1.03

[1] From LA Table 2420.B-10

[2] Note: ± values are only laboratory uncertainties (1σ) associated with the tritium value

[3] 1 foot = 0.3048 m

Second, this ± figure from the laboratory, unless specifically stated otherwise, represents only the uncertainty of the laboratory measurement, and does not include additional uncertainties associated with the field collection procedure itself (see, for example, Long and Kalin, 1990). Thus the true uncertainty of the measurement can be assessed only by repeated sampling and remeasurement of the same level. This may not be possible because in this case, as explained below, the sampling process itself very likely affects the system under study. It is important for the reader to bear in mind that the stated ± figures are minimal, and the true uncertainties (level of accuracy) were not evaluated. They are certainly greater than the reported values.

The procedure for collecting water from the unsaturated zone at the Ward Valley site involved the pumping of soil gas from air piezometers at specific depths through plastic tubing. The collected water vapor is then condensed in the solid phase in a cold trap at liquid nitrogen temperature (about -180°C). Collection continued for each sample at 1500 liters (l) of air per hour until 40 milliliters (mL) of liquid water were collected. This required 4 to 8 hours of collection. Thus 6,000 to 12,000 liters of soil gas passed through the collection apparatus for each water sample for tritium analysis (LA Appendix 2420.B, Figure 2420.B-6).

In view of the importance of the possibility of significant amounts of vertical movement of tritium (water) in the unsaturated zone, it is necessary to consider all possible explanations for the observed tritium data. Several possible explanations evident to the committee for the reported tritium found at depth in the soils at Ward Valley include:

1. <u>Infiltration of liquid-phase water within the past 40 years to a depth of at least 30 m</u>: The highest tritium values in the profiles (up to 6 TU) are the shallowest ones in the profile. Tritium values regularly decrease with increasing depth. Irregularities with this trend are not statistically significant. This pattern is consistent with liquid-phase vertical infiltration of precipitation containing moderately low values of tritium. The bomb pulse is not clearly apparent in these profiles, as its maximum would have decayed only to somewhat greater than 200 TU by the time these samples were taken and measured (1989). **Ages for water estimated from the chloride data indicate that under piston flow conditions the bomb tritium peak should be found in the upper 1 to 2 m of the soil. However, dispersion and mixing of the bomb-derived tritium-rich liquid water with tritium-free liquid water could have produced the observed tritium profile. The tritium profile could also represent very recent infiltration and dilution with tritium-free water that had been in the unsaturated zone for over 40 years. A finite level of tritium (>2 standard deviations above the detection limit) (See Box 3.2 for discussion of standard deviation) was reported at 30 m in boreholes GB-4 and GB-6. If this represents advective liquid movement in the unsaturated zone, it is inconsistent with the limited soil physics and chloride data at the site and with the applicant's conceptual model of water movement through the unsaturated zone.**

2. <u>Diffusion of gas-phase water through the unsaturated zone</u>: The license applicant proposes a vapor-phase diffusion model to explain the observed tritium profiles in the upper 9 m of the unsaturated zone (Harding Lawson Associates, 1990).

Modeling of the tritium found in boreholes GB-1, 4 and 6 contained in the license application and in a letter report (Harding Lawson Associates, 1990) does not account for the interaction of gas- and liquid-phase tritium, which has resulted in a significant overestimation of tritium migration. This was correctly pointed out in Wilshire et al. (1994). **The committee conducted modeling calculations, which included the effects of liquid interaction, and concluded that gaseous diffusion of tritium will be limited to only 1.5-3.0 m below the land surface at the Ward Valley site. Therefore, the vapor phase diffusion proposed in the license application as a mechanism to explain the observed tritium profiles does not explain the observations.**

Box 3.2

STANDARD DEVIATION

A standard deviation, abbreviated as σ (sigma), is a statistical statement of the precision of a measurement, assuming that all possible measurements of a quantity, for example the tritium content of a water sample, would have a "normal" or bell-shaped distribution about the true value. Because each water sample was measured only once, the numerical value of the standard deviation is a statement of the uncertainty of the measurement. In the case of tritium measurements, the value is based on the number of tritium decays measured by the analysis laboratory as well as on the repeat analysis of standards and laboratory blanks. The value does not, in the case of these tritium samples, represent variability and uncertainty in the field collection and processing.

The implication of this statistical phenomenon is that, in the case of the values that the laboratories reported for the standard deviation, for example, 10±1 (1σ), the probability is about 68% that the true value lies between 9 and 11. Correspondingly, for 2σ, or 2 standard deviations, the probability is about 95% that the true value lies between 8 and 12. Stated another way, the probability is about 5% that the true value lies outside of the 8 to 12 envelope. Still another statement from this example, the probability is about 2.5% that the true value is less than 8.

In the case of distinguishing finite levels of tritium from no measurable tritium, the 2σ criterion is used, giving one an approximate 95% confidence that the two normal curves (representing blank and sample) do not overlap.

More recent modeling to correct their earlier analysis (Harding Lawson Associates, 1990) that led to the vapor phase model concurs with the committee's findings and suggests "that only very limited downward transport of tritium occurs in the vapor phase". In addition, a fully coupled two-phase simulation was conducted that is a more correct and comprehensive interpretation of the diffusion process. The two-phase approach included the additional effects of liquid diffusion. Using this approach, the modeling results were reported to indicate "that some of the measured tritium concentrations in piezometers may be attributed to background levels ..." (Harding Lawson Associates, 1994b), where background refers to tritium which has moved by diffusion from the land surface.

The two-phase modeling approach is conceptually correct, although the liquid diffusion coefficients used in simulations are much larger than would be expected in the dry soils at Ward Valley. Two diffusion coefficients were used, 1.49 cm^2/day and 4.47 cm^2/day. These values (Harding Lawson Associates, 1990) were intended to represent the liquid diffusion coefficient of tritium in the soil, but the values represent the approximate, free-water diffusion coefficient and three times the free water diffusion coefficient, respectively, as reported by Gvirtzman and Margaritz (1986). It appears that an error was made in the Harding Lawson Associates letter report (1990), which calculated a free-water diffusion coefficient for tritium of 42.5 cm^2/day. In contrast, Gvirtzman and Margaritz (1986) suggested a free-water diffusion coefficient of 1.5 cm^2/day, a factor of almost 30 lower. Based on the same tortuosity as reported in the more recent Harding Lawson Associates letter report to the committee (1994b), the committee calculations resulted in an appropriate value for liquid-phase diffusion of tritium in the soil of 0.05 cm^2/day, much lower than that used in the two-phase modeling. Using the correct diffusion coefficients significantly reduces the enhancement of liquid diffusion in tritium transport. **As a result, it is virtually impossible that the tritium reportedly found in the boreholes can be explained ,by either single- or two-phase gas and liquid diffusion only.**

3. Barometric pumping: **Barometric and thermal pumping (Weeks et al., 1982) due to atmospheric conditions is inadequate to explain tritium at the observed depths, because wind and evaporative processes affect only the first meter or two below the surface (Weeks et al., 1982). Concentration-gradient driven diffusion is the mechanism of greatest importance for gas transport in the unsaturated zone.**

4. Artifact of collection procedure: The collection procedure itself could affect the level of tritium reported in the unsaturated zone. The large volumes of air removed from the unsaturated zone (up to 12,000 liters per sample taken over a period of up to 8 hours per sample) would remove air from the immediate vicinity of the sampling point, in homogeneous media, from a sphere surrounding the collection point. This sphere of extraction, in the case of the soil air sampled for tritium analysis at Ward Valley, would have a radius ranging from about 2 to 2.5 m, depending on the volume collected, and assuming an average porosity of 27 percent and an average water content of 7 percent in homogeneous media. The Ward Valley alluvium consists of variable mixtures of poorly sorted sand, silt, gravel and clay, as described in the license application. Inhomogeneity would distort the ideal sphere. Alluvial sediments tend to have greater textural variability vertically than horizontally. This would tend to flatten the ideal sphere and extend it horizontally. Textural variability in alluvium can also create preferential pathways in a vertical direction, but less commonly. Even microfissures and near-vertical cracks, which are unable to transmit water under unsaturated conditions, can easily conduct air under pressure differential. Microfissures, though not identified in the site characterization, are not uncommon in alluvial sediments. A third possible pathway is along the annulus of the borehole casing. This could occur if the space between the outer surface of the casing cylinder and the sediments is incompletely filled with grout. Such a possible leakage pathway need not extend to the ground surface, and thus could be concealed.

The grout itself is unlikely to be a source of tritium, as tritium-free water was used in its mixture. The air used during the "dry" drilling process contained about 6 TU. The relative and absolute humidity of the air in Ward Valley is normally very low. At 30° C and 10 percent relative humidity, the atmosphere at Ward Valley contains 3 mg/l of water. Therefore, some atmospheric tritium may have been transferred to depth during the drilling process, and this could have contributed to the non-zero tritium measurements observed. Were this an important addition to the tritium in the unsaturated zone, the second samplings from GB-4 would have showed significantly lower tritium than the first. This was not observed.

In summary, when soil-gas is extracted at depth from an air piezometer, the locality from which the gas is removed is indeterminate without complete knowledge of the textural variations and structural integrity of the sediments in the immediate vicinity of the sampling point, and of the degree of perfection of construction of the piezometers.

The low sampling yields (about 40 percent of the expected volumes of water) are consistent with mixing of the soil gas with low-humidity atmospheric air during sampling of the boreholes. Also consistent with this possibility is the fact that the highest values collected (5.6 and 6.0 TU at a depth of 5.1 m from GB-4) are only slightly lower than the measured tritium content of the ambient air (6.9 TU). The committee concludes that the possibility of atmospheric contamination of the samples in the uppermost levels of the unsaturated zone is likely, given the low levels of tritium observed and the low sample recovery. As discussed above, these results are also consistent with small amounts of recent infiltration to these shallow depths.

Atmospheric contamination may have affected the tritium measured at the collection points nearest the land surface (at 5 or 6 m), but it is difficult to apply these processes as possible explanations for tritium found in the deeper levels (11 to 30 m). Even considering variability in soil texture, the volumes of air in the sediments within a few meters of the piezometer sampling points are many orders of magnitude greater than the maximum volumes of air sampled for tritium analysis.

The only leakage pathway that could bring atmospheric air into the deeper air piezometers would be along the borehole casing. The committee considers this pathway to decrease in likelihood with increasing depth. The similar pattern of observed tritium in each borehole, namely decreasing with depth to, or almost to, the detection limit, is also consistent with each soil-gas sample representing the tritium level in the unsaturated zone at the collection depth.

It is thus not possible from the data available to conclude that the observations demonstrate infiltration during the latter half of the 20th century. The question then turns to the integrity of the tritium measurements on the deeper soil-gas samples: can the available tritium data allow us to distinguish between the two hypotheses for the measured values of tritium at or below 11 m in the unsaturated zone? The hypotheses are: (1) tritium, and thus meteoric water infiltrated to at least 30 m within the past 40 years, and (2) the reported tritium values lie within the analysis system blank, and therefore these values do not represent infiltration.

Procedural Blank and Uncertainty in the Blank

To verify the level of accuracy of the instruments and measurement procedures, it is necessary to obtain a procedural blank. Procedural blank determination in any analytical measurement is necessary to evaluate and interpret the significance of the result. This is done by using the instruments and procedures to measure a solution known to be free of the substance to be measured. False positive measurements then can be incorporated as the error values to be expected. The closer the expected value of the field measurement is to zero, the more critical is the blank determination.

All analytical procedures, especially those dealing with very low levels of the measured substance, require special attention to all steps in the procedure, beginning with the collection and handling, and ending with the analysis and reporting (Taylor, 1987; Long and Kalin, 1990). Laboratories typically report only the laboratory analysis portion of the uncertainty because that is the only series of steps they have control over. Table 3.1 gives the laboratory uncertainties. The report of the analyses in the license application (LA Section 2420.B, Table 2420.B-10) makes no reference to a procedural blank for assessing the true total precision of the analytical and collection procedures. Thus the total uncertainty (± figure) and blank[2] level for tritium for these analyses using the procedure described, are unknown. What is known is that the total analytical uncertainty, including collection, is either the same as reported in Table 3.1, or greater. In the committee's view, field collection procedures employed here may have added more uncertainty to the data, and the true uncertainty in the blank value, i.e. the real ± figure, is greater than reported in Table 3.1. For example, any joints in the sampling instruments must be vacuum-tight to prevent surface air from being entrained in the sample stream. To establish a collection procedure blank requires application of identical procedures to a hydrogeologic system for which one has prior knowledge of undetectable tritium. It is tempting to suggest that the 30-m sample from GB-1 is the value for the blank, and all others are finite. However, a single analysis cannot constitute sufficient information on which to base an estimation of the total analytical uncertainty and blank value. The true value for the tritium blank using this procedure remains unknown.

It is thus inappropriate to conclude from these data that the levels of tritium reported at or below 11 meters are either evidence for small amounts of post-1950 infiltration or simply within measurement uncertainty of zero. The tritium levels reported at depths above 11 m are consistent with small amounts of infiltration to those depths, but are also consistent with the samples that have incorporated some atmospheric water vapor during the sampling procedure.

Conclusions Regarding Tritium Found in the Unsaturated Zone

In the committee's judgement, the gaseous diffusion model presented in the license application as the most likely explanation for tritium at depth in the unsaturated zone is incorrect. Infiltration of meteoric water, as the Wilshire group suggested as the explanation of tritium at depth, is inconsistent with chloride and water potential data. In the committee's opinion, a likely explanation of the reported tritium at depth is sampling and procedural practices. The very low tritium values at depths below 11 m

---

[2] In analytical work involving measurements near the detection limit, the level of instrumental signal corresponding to true zero must be evaluated. This is referred to variously as background, noise level, or blank. This blank is subtracted from the initial instrumentation reading before reporting the results. This blank level has a built-in uncertainty, which here is expressed as a number associated with the ± symbol. Any change in analytical procedure or sampling conditions may affect the values of both the blank and the uncertainty figure. If the blank level has been underestimated, a measurement on the high end of the true blank can be misinterpreted as a positive reading.

reported in the license application are meaningless without the determination of a procedural blank for the method used to collect the samples. Thus these tritium measurements cannot be used to infer infiltration.

In light of plans to monitor tritium during site operation, the explanation for the reported tritium values should be resolved before the site monitoring begins. Other tracers, such as ^{36}Cl, have successfully been used elsewhere to constrain the magnitude of liquid water movement in the unsaturated zone and should be considered. Sampling for ^{36}Cl can be accomplished much more readily than for tritium at the Ward Valley site because large quantities of water are not required for analysis and chloride concentrations in the soil water are extremely high.

Conditions at Homer Wash

Features such as Homer Wash represent areas of likely higher fluxes and recharge. Periodic surface flow combined with coarse-textured soils at the surface and lack of extensive vegetation at the bottom of the wash provide an excellent opportunity for moisture to move below the zone of active evaporation and root extraction. Therefore, the observed lower resistivity zone beneath Homer Wash is most likely caused by higher water contents at depth. While the water content of the unsaturated zone may be relatively high, a continuous reflector detected in the seismic survey across Homer Wash and coincident with the water table indicates that the water table is not significantly higher beneath Homer Wash.

The committee agrees with both Wilshire et al. (1993) and the California Department of Health Services (1994) that active ground-water recharge may be occurring beneath Homer Wash. Inasmuch as Homer Wash is 760 m laterally from, and 10-15 m below, the base of the proposed disposal trenches, the committee concludes that such recharge is unlikely to have significant impact on water movement directly beneath the disposal site.

It should be pointed out, however, that although recharge from Homer Wash would not affect releases of radionuclides from the site nor water movement directly beneath the site, it probably would affect ground-water flow patterns in the saturated zone. One such effect might be to reduce or eliminate ground-water movement from the west, up-slope side of Homer Wash (where the site is located) eastward to the Colorado River, if such pathways actually exist, because of a possible hydraulic barrier effect of recharge from Homer Wash.

Summary of Subissue Conclusions

1. Based on the data presented in the license application and supporting documents, the committee concurs with the conclusion that water flux in the upper 30 m of the unsaturated zone is extremely low. This conclusion is based on the observations of low water content, low water potential, limited infiltration during the

monitoring period, and the significant accumulation of chloride in the upper 30 m of the unsaturated zone. Monitoring hydraulic parameters in dry desert soils is complicated. Monitoring instruments such as thermocouple psychrometers and heat dissipation probes are not robust and have a high failure rate. Although many problems were associated with the instrumentation at the site, particularly the monitored water potentials, all the water potentials monitored by the psychrometers fall within the dry range (-3 to -6 MPa). In these very dry systems, the level of uncertainty in hydraulic parameters that can be tolerated and still have little significance with respect to water fluxes is higher than in much wetter systems.

The committee concludes that monitoring of the unsaturated zone using corrected methodologies and equipment should have been continued after submission of the license application to provide a more complete data set. The committee also concludes that baseline data from a greater depth in the unsaturated zone should be collected as part of the monitoring program at the site.

2. Because of the extremely low water fluxes, it is difficult to resolve the direction and rate of water movement at the Ward Valley site. The license application states that the direction of water flow is upward; however, problems with the thermocouple psychrometers at the site made it difficult to determine the water potential gradient. In the committee's judgment, the most likely current direction of water movement at the Ward Valley site is upward. We based this on the low water potentials observed and the fact that similar sites in arid regions have vertical matric potential gradients in the unsaturated zone which drive the water vapor upward.

3. Based on model analysis, the committee concludes that the presence of tritium in the deeper portions of the unsaturated zone cannot result from vapor diffusion as stated in the license application. After consideration of several other possible explanations for the measured tritium, the committee concludes that the experimental design for collecting tritium from soil gas is seriously flawed, and this is the likely explanation for the observed tritium profiles below 11-m depth.

The committee also appreciates the difficulty of making applicable blank runs using the water-collection procedures used for the license application. Thus we recommend that future unsaturated-zone tritium measurements employ vacuum distillation collection procedures similar to those described in Yang (1992), or other direct water-isolation procedures. Any future attempt at collecting tritium samples must establish procedural blanks and evaluate potential contamination problems, so that a true positive value can be identified. We further recommend that tritium measurement techniques other than β-counting be investigated. Two such approaches, still under development, may enable more sensitive measurements using smaller samples of water. These include the ^{3}He-ingrowth method and accelerator mass spectrometry (AMS). These may be adaptable for unsaturated-zone monitoring. Additional analyses and data from tracers such as ^{36}Cl could help resolve this issue of unsaturated-zone infiltration.

4. The committee concludes that recharge at Homer Wash is possible but is unlikely to affect the ability of the site to isolate the wastes in the unsaturated zone.

EVIDENCE FOR RECHARGE TO THE GROUND WATER BENEATH THE SITE

Observational Methods

Direct observation of soil-water movement is generally impossible without disturbing the system. However, when soil water reaches the water table, this recharge can often be detected in the hydraulic and geochemical response of the uppermost part of the ground-water flow system. Such response could include the appearance of a contaminant or tracer from the surface in a monitoring well, changes in the ground-water chemistry, rises in the water table following a recharge event, or by a vertical hydraulic gradient in the aquifer. In general, these techniques do not provide quantitative evidence of flow, but rather are useful qualitative indicators.

Tracers commonly used in arid zone ground-water studies are similar to those used in the unsaturated zone and include tritium (^{3}H) (indicative of recent recharge), ^{14}C (indicative of the approximate age of the ground water) and the stable isotopes of water, deuterium (^{2}H) and ^{18}O (to indicate climatic conditions during recharge). The Wilshire group drew attention in particular to the ^{14}C age dating technique used in the license application, which concludes that the ground water is very old and could not have been recharged under the present-day climate (Wilshire, 1993a,b, 1994). In addition to these tracers, the committee has reviewed other indicators of recharge at the water table that were discussed in the public meetings of the committee or have come to light in our review of data.

Tritium in the Ground Water

As a chemical tracer in hydrologic studies, the presence of tritium (>0.2 TU) in ground water indicates recharge during the past 40 years. The 12.3 year half-life of tritium makes it ideal for tracing young waters or identifying young (<40-year old) components in mixtures. This makes it possible to determine if ground waters are less than 40 years old, or if they have had significant input of meteoric waters during the past 40 years.

Water samples from zones 5 to 68 m below the water table were taken from the monitoring wells in order to establish whether recent infiltration had reached that level. If tritium were found above background levels in these samples, it would be positive evidence for infiltration of water during the previous 40 years. With one exception, all ground-water samples were within the limit of detection of the laboratory analysis (1 TU). The exception, 3.7 ± 1.0 TU, was one of two duplicate samples from well WV-MW-04. The other duplicate sample, collected on the same day, yielded 0.0 ± 1.0 TU. The license application states that the finite analysis is spurious in light of the fact that the duplicate and all other tritium samples from the saturated zone yielded no measurable tritium. This statement is corroborated by freon measurements. Freons, which are man-made gases, are present in ground waters recharged in the second half of this century. Freon determinations were negative in all water samples from the monitoring wells at the Ward Valley site.

An alternative point of view was presented in one of the committee's open meetings (Committee to Bridge the Gap, 1994) suggesting that more sensitive tritium measurement technologies should have been used to enable detection of lower proportions of recent recharge in the ground water. However, due to the possibility of *in situ* tritium production, measurements ≤ 0.2 TU would have ambiguous interpretations (Lehmann et al., 1993).

Conclusions Regarding the Observation of Tritium in the Ground Water

In light of other tritium analyses on samples taken from the monitoring wells, all within about 1000 m of each other, that showed no measurable tritium, the committee concludes that the one 3.7 TU value is likely in error. In any low-level radioactivity measurement, a false positive is more likely than a false negative. The tritium data, therefore, do not indicate recent recharge to the ground water at the site.

Carbon-14 Age of the Ground Water

^{14}C has been widely used to date ground water, although ground-water ^{14}C analyses are not straightforward enough to interpret in terms of time since recharge. The results presented in the license application were on water collected from monitoring wells penetrating the same hydrologic unit, and located within about 1000 m of each other. These ages, with no model-based adjustments, range from $12,560 \pm 750$ to $16,920 \pm 590$ years (LA Appendix 2600.A). Subsequent analyses provided to the committee in a report (Grant Environmental, 1994) revised these age estimates according to several model age-adjustment protocols to range from 4,500 years to 22,000 years.

The ^{14}C data on ground water, taken from the dissolved inorganic carbon (DIC), can be interpreted only in the context of hydrogeochemical models. As the available data do not unequivocally reveal recharge areas or even flow directions for Ward Valley, we discuss the data more generally in terms of a likely hydrogeochemical scenario. A viable general model for ground-water flow and geochemical interaction for ground-water alluvial aquifers in arid and semi-arid basin-and-range valleys involves: (1) recharge near mountain fronts, (2) possible recharge beneath drainageways (streams and washes), (3) possible recharge through alluvium, (4) chemical interaction between ground water and aquifer minerals, and (5) gas/liquid-phase interaction between water in the saturated zone and gases in the unsaturated zone. In the context of this general model, one can evaluate the likelihood of these processes, in addition to radioactive decay, affecting the ratio of ^{14}C to total measured carbon ($^{14}C/C_{TOT}$, usually expressed as percent modern carbon, pMC) in the DIC.

1. Mountain-front recharge water derives its DIC from dissolution of soil gas CO_2, which has a $^{14}C/C_{TOT}$ equal to or nearly equal to that of the atmosphere (Fontes, 1992; Kalin, 1994). Some dissolution of calcite may occur in this environment, which would reduce the

$^{14}C/C_{TOT}$, but isotopic exchange with gas-phase CO_2 from plant respiration tends to return the $^{14}C/C_{TOT}$ in DIC back to modern levels.

2.   Recharge at drainageways, like Homer Wash, through the unsaturated zone to the water table would act as a short circuit and mix possibly younger water with older water in the saturated zone. The resulting mixture would have a higher $^{14}C/C_{TOT}$ (higher pMC) of the DIC, thus decreasing the apparent age.

3.   Recharge through the alluvium would have the same effect as process 2 on the apparent ground-water age.

4.   Chemical reactions in the aquifer either have no effect on the apparent ^{14}C age (for example, incongruent dissolution of silicates or precipitation of calcite), or increase the apparent age (for example, dissolution of calcite or oxidation of lignite). These reactions, and their effects on the $^{14}C/C_{TOT}$, can be modeled, but only if a downgradient sequence of water samples is available. NETPATH (Plummer et al., 1991) was designed for this purpose.

5. $^{14}C/C_{TOT}$ of DIC in the saturated zone could be affected by dissolution of, or exchange with, CO_2 in the unsaturated zone. This process could operate throughout the entire basin. Fontes (1992) stated that this would be most pronounced in cases of high pH ground water. Kalin (1994), who modeled reaction paths in the Tucson Basin in Arizona using NETPATH, did not find CO_2 exchange to be important in the Tucson Basin. In the case of the Tucson Basin, the samples were from wells used for water supply and were completed deeper below the water table than the monitoring wells at Ward Valley. WATEQ (a subroutine in NETPATH) analyses of the ground water from the Ward Valley monitoring wells (Grant Environmental, 1994) revealed that all water samples were essentially in equilibrium with calcite. This suggests that calcite dissolution may have affected and (probably decreased) the $^{14}C/C_{TOT}$ of the DIC for the Ward Valley samples.

Because of the overall lack of hydrogeochemical constraints on the processes discussed above, including large uncertainties in $\delta^{13}C$ and $^{14}C/C_{TOT}$ values in soil-zone CO_2 and in aquifer carbonates, and because of the possibilities of DIC $^{14}C/C_{TOT}$ alteration by processes discussed above, we cannot place confidence in any specific ground-water age. However, if significant surface water had infiltrated in the vicinity of the Ward Valley proposed site within the past few thousand years, but before the nuclear weapons-induced pulse in 1960, the water-table surface would reveal ^{14}C levels in the 50 to 80 percent modern carbon (pMC) range. This results from reactions of DIC with aquifer carbonates. Pre-1950 atmospheric CO_2 and living plant tissue had ^{14}C concentrations of 100 pMC; aquifer carbonates can range from 0 to 100 pMC. Measured values of ^{14}C in the water sampled from the monitoring wells ranged from 11 to 22 pMC (Grant Environmental, 1994).

Because processes other than radioactive decay can affect the apparent radiocarbon age of ground water, ^{14}C may not be an ideal tracer of recent recharge. Exceptions to this are systems that undergo insufficient carbon-diluting processes that may attenuate the "bomb pulse" to levels below 100 pMC. In other words, the following conditions must be met for ^{14}C in ground water to be useful in identifying vertical recharge during the second half of this century: water infiltrated from the surface near the proposed site vertically to the water table, all within the 1958 to 1989 time frame, without dissolving enough soil/unsaturated-zone carbonate or without mixing with enough existing saturated- or unsaturated-zone water to lower the bomb pulse to <100 pMC.

Ground waters that are chemically or isotopically stratified are difficult to sample without mixing waters of different ages and/or sources. This likelihood complicates the meaning of "ground-water age." For example, waters sampled from the monitoring wells were screened over a vertical distance of 14 m. The upper limit of these screens ranged from 5 m to 68 m below the water table. Mixing of waters of different ages and possibly different sources (long versus short travel times) is often a possibility. Consequently, in the case of mixtures, an infinite number of water mixtures of different "ages" is possible. A small proportion of "post-bomb" carbon could be concealed in a composite water with measured pMC values of 22 or even 11. Therefore, it is often impractical to calculate or assign a precise age to ground water. This is true for the Ward Valley data.

The two adjacent monitoring wells (WV-MW-01 and WV-MW-02) are screened from 5-17 m and from 60-70 m below the water-table surface, respectively. Two rounds of sampling were conducted on these wells. In the first round, samples from these two wells were not significantly different in $^{14}C/C_{TOT}$ (MW-01: 10.9 pMC; MW-02: 11.3 pMC), which suggest no vertical gradient in ^{14}C. However, the major-element chemistry reveals marked differences between these two waters. The Round 2 sample, however, from MW-01 (17.5 pMC) was significantly higher than the Round 1 sample (10.9 pMC). This apparent vertical stratification, whereby the upper portion of the aquifer was higher in ^{14}C than the lower, manifested itself only under pumping conditions. This has been seen elsewhere (Mazor, 1991) and suggests that the waters at the top of the water table were recharged more recently than deeper waters. As the Ward Valley wells were not screened slightly below the water table, it is not possible from the data to date this uppermost water.

Based on the information available, the committee finds it is not possible to rule out any of the ages presented in the license application, or to rank them in terms of relative likelihood. A fair statement is that the mean age of the ground water sampled beneath the Ward Valley proposed site is covered by the range of model ages presented in both the license application and subsequent documents; i.e. between 4,500 and 22,000 yr. However, because of the possibility of the water samples being mixtures of different ages, the Ward Valley ^{14}C data cannot unambiguously demonstrate the presence or absence of modern recharge. A Holocene or Pleistocene age of ground water below Ward Valley does not necessarily indicate local recharge, because ground-water flow paths are very long and much of the ground water presently beneath the Ward Valley site probably originated from recharge in higher elevation regions.

Stable Isotopic Composition of Ground Water as an Indicator of its Age

The stable hydrogen and oxygen isotopic composition of water can give important information about the origin of water. In particular, several important generalizations can be made about the origin of surface water based on the contents of deuterium (D or ^{2}H), a heavy isotope of hydrogen (H), and of oxygen (^{18}O) isotopes of such waters, where δD is:

$$\delta D(‰) = \left[\frac{(D/H)_{SAMPLE}}{(D/H)_{SMOW}} - 1 \right] \times 1{,}000 \qquad \text{(Equation 3.4)}$$

and H is the light isotope (^{1}H) of hydrogen, D is the heavy isotope (^{2}H) of hydrogen, and SMOW is the international isotopic standard for water (Standard Mean Ocean Water). The equations defining δD, δ^{18}O, and δ^{13}C are analogous. The units used are deviations from the SMOW reference for δD and δ^{18}O, and from the PDB reference in the case of ^{13}C . The values represent parts-per-thousand (‰) deviation from the respective stable isotope ratio standards.

Three types of water, with characteristic δD/δ^{18}O patterns, are important in this discussion: (1) Waters in which δD and δ^{18}O values lie on the meteoric water line (MWL), so called because most precipitation falls very close to the line (Figure 3.5), and can be defined as

$$\delta D = 8\ \delta^{18}O + 10\ \text{(Craig, 1963)} \qquad \text{(Equation 3.5)}$$

This is an empirical relationship. Each watershed could have its local meteoric line with characteristic slope and intercept. (2) Waters that are relatively enriched in δD and δ^{18}O due to evaporation fall on a trend with a δD/δ^{18}O slope less than 8 and between 3 to 5 and intersect the MWL. (3) Waters that have undergone extensive water-rock interaction and are enriched only in δ^{18}O (Figure 3.5). Normally, extensive water-rock interaction is restricted to geothermal waters.

Ward Valley Stable Isotope Data

The data in the license application come from five ground-water wells, each of which was sampled on two dates (January-February 1989 and May-June 1989). The reproducibility of the δ^{18}O of water samples does not seem to be as good as is quoted, especially for replicate samples analyzed by different laboratories. The committee considers that the data from WV-MW-02,03,04,05 are not distinguishable from each other and can be considered to be a single point. WV-MW-01 (the sample with the most positive δD value for both sampling dates) is,

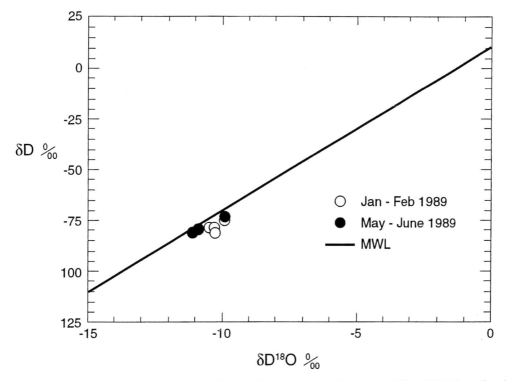

Figure 3.5   The isotopic composition of well water from the Ward Valley site in 1989 as reported by the Alberta Environmental Center. The solid line shows the meteoric water line (MWL) of Craig (1963). All samples plot very close to the MWL and are not distinguishable from meteoric water from southeastern California (collected by Smith et al., 1992).

however, distinguishable from the other water samples. It is also the sample closest to the water table.

The samples plotted in Figure 3.5 are very close to the meteoric water line, indicating the isotopic composition of non-evaporated waters in the region. The δD and $\delta^{18}O$ values are about -78 and -10.8 per mil, respectively, which is in the cluster of values found by Friedman et al. (1992) and Smith et al. (1992) for modern meteoric waters in the region. This δD value is similar to that of winter precipitation near Needles, CA, as shown in Figure 5b of Friedman et al. (1992) and is significantly higher than summer or mean δD values for the region.

The entire range of δD and $\delta^{18}O$ for all five wells is 7.0 per mil for δD and 1.1 per mil for $\delta^{18}O$ (using only data measured in the laboratory at Alberta Environmental Center (AEC). This small range, especially when the samples are so close to the meteoric water line, is not enough to determine if the waters had undergone significant evaporation after falling as rain.

Unfortunately, no "blind" duplicates were run in the Alberta laboratory, so that the true uncertainty is not known. The single sample that can be clearly identified as being a duplicate is from monitoring well WV-MW-05 analyzed by Global Geochemistry Corporation (GGC). The $\delta^{18}O$ value for that sample differed by 0.65 per mil between the two laboratories (in the original data sets, WV-MW-05 4 Feb 89 is -10.45 by AEC, and -9.8 [average of -9.9 and -9.7] by GGC). The uncertainty in the true accuracy and precision of the $\delta^{18}O$ measurement is in doubt, but given the paucity of duplicate analyses, it is certainly higher than the quoted uncertainty of ± 0.2 per mil.

Three samples (WV-MW-03,04,05) yielded $\delta^{18}O$ values that were 0.4 to 0.5 per mil depleted in the May 1989 sampling compared to the February 1989 sampling. This slope is characteristic of waters with high water/rock ratios, such as in geothermal systems, although the direction of the depletion is wrong since the sample should become more enriched through time. It is not characteristic of evaporation. However, it is most likely that these differences are within the true uncertainty estimates for this data set as shown by the differences in $\delta^{18}O$ between different laboratories on the same sample and later samplings of the same wells.

Estimates of Paleotemperatures Using the Claussen Approach

The license application proposed that the stable isotopic composition of the ground water can be used to infer the average annual air temperature at the time of recharge as proposed by Claussen (1986). **In the committee's judgment, the Claussen (1986) approach to estimating the paleotemperature of recharge is inappropriate for this study.** Claussen (1986) used δD and $\delta^{18}O$ values for monthly mean temperatures at a single mountain site in Colorado, where climatological conditions are vastly different from those at Ward Valley. Compilations of δD, $\delta^{18}O$, and temperature data by the International Atomic Energy Agency (IAEA) (1981) and discussions of such data by Rozanski et al. (1992), show that the $\Delta\delta^{18}O/\Delta T$ slopes are quite variable, and that the intercepts vary significantly even for stations with similar meteorology (e.g. Chicago, Illinois, and Ottawa, Canada) resulting in distinctly different recharge temperature estimates using the Claussen (1986) approach. In addition, site-specific variables such as differential infiltration of summer versus winter rains, as well as differences in storm-track patterns between glacial and interglacial periods, add even greater uncertainty to this type of calculation. The local meteorology of the Mojave Desert Ward Valley region is completely different from the mountainous Colorado site. The IAEA (1981) data for North American non-coastal sites using the Claussen (1986) approach give a range of inferred recharge temperatures for Ward Valley extending over 50°C (90°F). These calculated temperatures illustrate the futility of the Claussen (1986) approach (Table 3.2).

Conclusions Regarding the Stable Isotopic Signature of Ground Waters

In the committee's view, the δD and $\delta^{18}O$ data show that the isotopic composition of ground water in the region is indistinguishable from local meteoric water or from Holocene water. They do not provide an estimate of recharge temperature, nor do they provide any evidence of post-precipitation evaporation.

Table 3.2   Regression data and calculated recharge temperature at Ward Valley. Data from Claussen (1986) and the International Atomic Energy Agency (1981).

Location	Regression Equation	Recharge Temperature* (°C)
Claussen (1986)	$\delta^{18}O$ = 0.57 T - 15	7.0°
Chicago (IAEA)	$\delta^{18}O$ = 0.35 T - 10.62	-1.1°
Flagstaff (IAEA)	$\delta^{18}O$ = 0.34 T - 9.80	-3.5°
Ottawa (IAEA)	$\delta^{18}O$ = 0.32 T - 13.25	7.0°
The Pas (IAEA)	$\delta^{18}O$ = 0.40 T - 19.2	20.5°
Waco (IAEA)	$\delta^{18}O$ = 0.14 T - 6.37	-33.1°
IAEA (global MAT)	$\delta^{18}O$ = 0.56 T - 12.7	3.0°

* Estimated paleotemperature of recharge at the Ward Valley site, calculated assuming a $\delta^{18}O$ value of -11 per mil

Water Chemistry Changes

Although not commonly used to measure recharge, observed changes in ground-water chemistry can indicate recharge through the unsaturated zone. In recent years, monitoring ground-water quality has become routine at many industrial sites to detect if ground water has become contaminated by surface activities. If ground-water movement is slow compared to recharge velocities, chemical changes in the upper portion of the aquifer may be the result of recharge.

The geochemical composition of ground water is determined in large part by the mineral dissolution reactions that contribute ions to the water system. Dilution and mixing with other ground water are means to change significantly and quickly the chemistry of ground water. For certain ions, anthropogenic input can significantly alter their concentrations (e.g., notably Na^+ and Cl^-). Other ions are considered to be nutrients and are closely related to biological activities at the earth's surface. These include nitrate (NO_3^-) which can also be greatly affected by human activity. Other species, notably silica, are generally well buffered in ground water and change their concentrations due to dissolution of silicate minerals (although not quartz). Silica is not generally affected by human activity.

Chemical Variations in Ward Valley Ground Water

The license application and a letter report (Harding Lawson Associates, 1994a) conclude that the chemical variations are not the result of recharge events to the water table.

The letter report also concludes that the lack of seasonal trends in chemical variability can be used to support the hypothesis of no current recharge. In a deep unsaturated zone where precipitation is highly variable, seasonal response of the water table is, however, not expected.

The geochemistry of ground water from monitoring wells WV-MW-01 to -05 was reported in the original site license application and supplemented with data contained in a letter report (Harding Lawson Associates, 1994a). Although sufficient to discuss general aspects of the regional water chemistry, the data are not precise enough to determine if small changes occurred in water composition through time. Special care must be taken in sampling and chemical analysis to determine small changes in water composition through time. In addition, much higher precision would be needed than is normally provided by commercial laboratories.

The monitoring well chemistry shows considerable variation in certain chemical constituents (silica, total dissolved solids, nitrate, and bicarbonate), electrical conductivity, and temperature. Some portion of the variability can be ascribed to analytical precision in the laboratory, which is to be expected, or to field measurement errors. Duplicate samples sent to several different laboratories show that agreement for some species is very poor among laboratories. In the case of temperature, some of the variation was caused by poor sampling design.

The committee noted that nitrate values are reported as ppm N-NO$_3$ (Molecular Weight (MW) = 14, which is the MW for Nitrogen alone) for the earlier sampling, and are erroneously reported as ppm N-NO$_3$ for the May sampling but are actually ppm NO$_3$ (MW=62, the MW for the nitrate species (LA, Section 2600, Table 2600-1)). In Addendum 2600A.A, in which the May samples differ from Table 2600-1 by the factor 14/62, all the data are apparently reported in the same units (ppm N-NO$_3$). In this case, the problem between reporting concentrations would be avoided by using molar concentrations rather than units of weight. **The committee concludes that the nitrate data do not indicate local recharge as has been proposed by Wilshire et al. (1994), but rather indicate discrepancies in the reporting of laboratory data.**

Significant changes in silica concentration without notable changes in other dissolved components are very difficult to explain, because only significant dissolution of quartz or precipitation of quartz with no other changes in the chemical solution can cause this to happen. This is extremely unlikely in any natural system.

Variations in Specific Conductivity

Specific conductivity is directly related to total ionic strength. Reported measurements of changes in specific conductivity (by up to 50% of the average value) were not accompanied by changes in the major ion chemistry of the waters. This seems strange because the specific conductivity of a solution cannot be changed significantly without also changing the major ion concentrations. Since the changes in ion conductivity were synchronous with collection/analysis dates in spite of small or no changes in ion concentrations, the problem is most likely in the laboratory calibration. **The committee has,** therefore, concluded that the reported changes in specific conductivity do not provide evidence for current local recharge but rather are the result of sampling, analytical, or procedural errors.

Conclusions Regarding the Fluctuations in Water Chemistry

Chemical analyses of ground water below the Ward Valley site reveal that composition of the waters was derived from the weathering of intrabasinal sediments, common in the western United States. Ground-water chemical fluctuations are not expected in deep ground waters isolated from the land surface and infiltration. The data on the ground-water chemistry are of insufficient quality to determine if small temporal changes occur, for which a specific sampling program would have to be designed. Most, or all, of the reported temporal differences in the chemical composition of the Ward Valley site ground water are likely due to collection and laboratory procedural uncertainties and errors. **The committee concludes, therefore, that the reported ground-water chemical fluctuations are not attributable to current recharge, but rather are the result of sampling, analytical, or reporting uncertainties and/or errors.**

Water-Level Fluctuations

Numerous workers (e.g., Freeze and Cherry, 1979; Sophocleous and Perry, 1984) have reported rises in ground-water levels to infer recharge. The response of the water table to recharge is a function of depth to water, type of recharge, and soil and aquifer materials. Water-level rises can be rapid in the case of preferential flow or seasonal as is typically found in more humid climates with shallow water tables. Fluctuations in water levels have also been associated with flood events in ephemeral washes. Because, in most cases, it is difficult to ascertain how much recharge has occurred, the rises are qualitative indicators of recharge.

Numerous other factors besides recharge can result in water-level rises (and falls) and must be evaluated as alternative hypotheses when water-level changes are observed. These are summarized in Table 6.2 of Freeze and Cherry (1979). Those most likely to be encountered near the Ward Valley site include atmospheric pressure changes, ground-water pumping, and earthquakes. An additional source of observed water-level changes results from measurement error. Particularly in cases where the depth to water is great, small errors, changes in measurement technique, or inexperience can result in false indications of water-level changes.

Water-Level Measurements at the Ward Valley Site

Water-level measurements in the deep monitoring wells at the Ward Valley site have shown variation in depth to water during and subsequent to the monitoring period. The fluctuations do not appear to be correlated with either nearby wells or subsequent measurements. If the water table were responding to recharge events, we could expect a rapid rise in the water level, which should be followed by a slow decline in water level over time. Instead, the observed water levels generally showed rapid declines rather than the increases that would be expected under locally-derived recharge. Moreover, the variations are small (generally less than one meter) when compared to the depth of measurement and are likely to be within the range of error of the measurement technique.

Conclusions Regarding Water-Level Fluctuations

The committee concludes that the variations in water-level measurements reported for the Ward Valley site do not constitute evidence for local recharge and are most probably the result of measurement errors and uncertainties.

Ground-Water Gradients

In simple aquifer systems where flow is predominantly in a lateral direction, it is uncommon to encounter appreciable water-level differences in adjacent wells completed at different depths. Vertical gradients are commonly observed in recharging or discharging portions of aquifers where the flow direction is not primarily horizontal. In recharging areas, the flow lines below the water table dip downward, reflecting the accumulation of water (recharge) from the surface. At discharging areas such as at Danby Dry Lake, hydraulic heads are likely to increase with depth, supporting the discharge of water by evaporation. Between the recharge and discharging areas of large regional aquifers where the flow is primarily horizontal, water levels will generally be uniform with depth.

Monitoring wells WV-MW-01 and WV-MW-02 show an apparent downward hydraulic gradient of approximately 0.01 m/m. The process or processes responsible for this gradient have not been explained in the license application. If the apparent gradient is accurate, local recharge or discharge into a deeper zone are possible explanations for its existence.

A plausible explanation for the observed gradient may lie in the measurement techniques, however. For example, any deviations from the vertical in the boreholes could lead to erroneous depth measurements because the measurement of depth to the water table would not represent the actual depth, but the distance down the borehole to the water table. Such errors would be consistent throughout the monitoring period and could lead to an apparent difference in water level elevation. Deviations from the vertical could also mean that the vertical hydraulic gradient is much greater than the one determined, for instance, if MW-01 deviates from the vertical substantially but MW-02 does not.

Fluctuations in the gradient between the two wells were also observed in the reported data during 1989, 1991 and 1993. These apparent fluctuations do not appear

to be attributable to local recharge because the fluctuations were the result of reported declines in water level in well MW-02. If recharge were occurring at these dates, we would expect to see an increase in water level in MW-01 rather than a decline in water level in MW-02. During the monitoring period, procedural changes were made to the measurement method, including changes to the reference measurement point. These changes along with measurement uncertainty are the likely source of the reported fluctuations.

Conclusions Regarding the Observed Vertical Gradient

The source of the reported vertical gradient between wells MW-01 and MW-02 cannot be resolved with the data in the license application. The committee concludes, however, that the fluctuations in the gradient cannot be attributed to local recharge.

SUMMARY OF CONCLUSIONS OF SUBISSUES 3, 4, AND 5

1. The tritium isotope compositional pattern of the ground water strongly supports the license application conclusion that significant recharge to the water table is not occurring directly beneath the Ward Valley site. However, the uppermost saturation may not have been sampled, and some ^{14}C data indicate stratification may exist at the site in the saturated zone.

2. The ^{14}C age estimates in the license application may not represent minimum ages for the ground water, based on the committee's reevaluation of the ^{14}C and stable isotopic data. Rather, the data suggest a minimum age of 4,500 years for bulk water samples from beneath the site.

3. The variations in ground-water chemistry are not attributable to ground-water recharge. Most can be shown to be the result of procedural and laboratory errors.

4. Apparent fluctuations of the water table during the monitoring period do not indicate recent recharge. Most are likely to be the result of procedural errors and the precision of the measurements.

5. The cause of the observed vertical gradient in monitoring wells WV-MW-01 and WV-MW-02 cannot be resolved at this time, but the fluctuations in the gradient cannot be attributed to local recharge.

6. Data on tritium, ^{14}C, chemical composition, and water level fluctuations of the ground water at Ward Valley do not support recent recharge events.

ADEQUACY OF THE PERFORMANCE-ASSESSMENT MODELING

Wilshire et al. (1993b and 1994) concluded that the data collection efforts and modeling do not adequately represent the complexities of the processes of water movement in the unsaturated zone at the Ward Valley site. The DHS has stated in its reply (DHS, 1994) that the data are sufficient and the performance modeling is conservative. This conservatism is sufficient to include any uncertainties in the site characterization data brought about by the complexities of the site.

The committee assessed the adequacy of the data to conduct performance-assessment modeling. Based on that assessment, the committee then evaluated the performance modeling with special attention to the conservative aspect of the modeling.

Adequacy of the Database for Performance Modeling

Modeling soil water and contaminant transport in arid soils is a relatively new field. The data requirements for such modeling are quite daunting and often difficult or impossible to obtain with the available technology. Assumptions must be made at all stages of modeling on the relative dominance of processes, such as heterogeneity, sorption phenomena, and boundary conditions. Assumptions must also be made regarding specific properties, such as hydraulic conductivity, when such data are not available. When data are uncertain or not available, a range of the expected values can be used to bracket the behavior of a system and to determine the sensitivity of the outcome to the assumptions.

Data needed to model water and solute transport fall into two major categories, the boundary conditions, and the soil properties. Boundary conditions include the precipitation, its timing and intensity, the evapotranspiration at the land surface, the depth to the water table, and the concentration of radionuclides in the disposal trenches. Soil properties include, but are not limited to, the hydraulic conductivity, the spatial distribution of conductivity in the soil, and the sorptive properties of the soil. Many of these data are very difficult to measure or cannot be measured at all locations in the soil. The performance modeler's responsibility is to estimate these data where unavailable in a manner to (a) insure that the model is conservative in its prediction, and (b) insure that the estimates are reasonable, given the understanding of the physical system.

One of the most difficult problems facing modelers is the complexity and spatial distribution of hydraulic properties in the soil. Numerous studies have shown that soil properties vary considerably across the landscape and in the vertical direction. These are also some of the most difficult data to obtain. Butters et al. (1989), among other studies, demonstrated that soil heterogeneities strongly affect solute movement at different scales. However, few studies have focused on arid systems. Cook et al. (1989) found that soil-water flux, or percolation, in Australia was spatially variable and controlled by both topography and soil texture. In contrast, Hills et al. (1991) observed that even in heterogeneous soils, the assumption of uniformity appears to yield satisfactory results (to first order) with respect to infiltration into dry soils. Resolution of the role of heterogeneity in arid systems is still an emerging science and it is not yet clear how important it will be in the long-term disposal of radioactive waste. It is, therefore, critical that modeling be conservative when such data are not available.

While heterogeneity of the soil profile will clearly control the distribution of water and solute flux, more important to modeling arid soils is understanding the boundary conditions. In arid systems, the availability of water at the land surface is critical in controlling the rate of deep percolation and recharge. If water is not available from the land surface, little water or solute movement will occur. It is therefore critical to quantify the availability of water and the efficiency of plants and evaporation to remove water from the near surface. Fortunately, these data are often easier to obtain.

Water and solute modeling in the unsaturated zone at Ward Valley were conducted to simulate a base case and hypothetical failure modes of the trenches and covers. The modeling used both site-specific data as well as reported properties from areas judged to be similar. In the modeled cases, the soils were assumed to be homogeneous and their characteristic properties derived from either the literature, measurements on core samples, or from calibration with field experiments. The soil core data indicated variability in hydraulic conductivity. However the infiltration test data do provide an averaged estimate of the *in situ* hydraulic properties. In all cases, these data were obtained from the upper 30 m of the unsaturated zone. Other than drilling logs, no data were available below a depth of 30 m to the water table (at approximately 182 m).

From a modeling perspective, the choice of homogeneity in soil properties is not unreasonable, given the large scale of the problem and difficulty in obtaining data. Extensive use of the infiltration test data and calibration is also justified in light of these difficulties. These data appear to be consistent with literature values, although the departure of the calibrated hydraulic conductivity data at low water contents that was found in the infiltration experiment is not well explained in light of other literature results. This departure appears to be an artifact of the model calibration and does not have any justifiable physical significance. Its importance to the performance-assessment modeling appears to be minor.

The lack of data deeper in the unsaturated zone is of more concern. While the coring data showed similar lithologies at depth, few hydraulic data were obtained from these depths. As the waste trenches may extend to a depth of 18.5 m, much of the unsaturated zone at the site has not been hydraulically characterized.

The upper boundary conditions, rainfall, and evapotranspiration used in the simulations are based on local measurements and literature data. Average annual precipitation data were used for much of the base case modeling. Intensity data were not available, but extreme events were used for the failure cases.

Conclusions Regarding the Data for Modeling

In summarizing the adequacy of the soil data to quantify the heterogeneity, the committee recognizes the difficulties in obtaining such data. **The modeling of the unsaturated zone using the calibrated data from the infiltration test is justified. In the committee's opinion, additional data from deeper in the unsaturated zone than the 30**

meters provided by the license application would provide greater confidence in the modeling results.

In the committee's opinion, the use of average annual precipitation and estimated intensity data for the base cases is not conservative although its effects on the base case are unlikely to affect performance significantly. The use of extreme events for the failure scenarios is, however, conservative and justified.

Modeling of the Complex System

The committee examined the performance modeling in light of the available data to ensure that it represents likely and credible scenarios and provides conservative bounds on the performance of the site.

A wide variety of performance objectives were modeled in various scenarios (Harding Lawson Associates, 1994c), ranging from an undisturbed unsaturated zone under ambient climate to a simulated failed B/C trench cover. In all cases, the soils were assumed to be uniform from the land surface to the water table. Scenario modeling was also conducted for the saturated zone and for vapor transport. These latter efforts do not, however, directly pertain to the issue of the potential for transport through the unsaturated zone. The boundary conditions and complete summaries of the simulations are found in Harding Lawson Associates (1994c).

Scenario 1: Undisturbed Unsaturated Zone Under Normal Climatic Conditions

The purpose of this effort was to determine if recharge could be occurring presently at the site under undisturbed, i.e. vegetated, conditions. Both soil properties from the literature on similarly textured soils and vegetation data for selected species at the site were used. Climatic data were estimated, including annual precipitation of about 12 cm. The unsaturated zone was assumed to be homogeneous and isotropic in its properties.

The results of the modeling revealed no net downward movement of water below the active root zone during the simulated twenty years. These results support the conclusion that transport by liquid flow through the unsaturated zone is negligible for the conditions modeled.

No above-average rainfall years were simulated. Although use of increased rainfall would appear to be conservative initially, as higher rainfall years should allow deeper infiltration of moisture, it is unlikely that recharge would occur even in wetter years, given the efficiency of desert vegetation and its fast response to changes in annual precipitation.

In the opinion of the committee, the assumptions used in the modeling are reasonable for the Ward Valley site. The use of non-site hydraulic data is less than optimal, but should have little impact on this scenario because of the strong control of the water budget by the vegetation. Under normal desert climatic conditions, the vegetation, as an excellent scavenger of water, will leave little water for recharge.

Scenario 2: Undisturbed Unsaturated Zone Without Vegetation

This scenario was modeled to determine if recharge would occur under non-vegetated conditions. Studies have shown (Gee et al., 1994) that recharge can occur in arid climates if vegetation is absent. The hydraulic and climatic conditions were identical to those used in the previous scenario.

The results of this modeling of the Ward Valley site showed that recharge would occur at a rate of 0.35 cm/yr if vegetation were removed from the soil surface. This represents approximately 3 percent of average annual precipitation and is considerably less than that reported in other studies. Gee et al. (1994) reported significant increases in soil moisture and recharge when vegetation is absent at an arid site in eastern Washington. Recharge in excess of 50 percent of the precipitation was reported when only shallow-rooted grasses were present at this site (Gee et al., 1994). Gee et al. (1994) have shown increases in water content beneath unvegetated trench covers at the Beatty, Nevada, low-level radioactive waste disposal facility. No flux values were reported, but the increases in water content were small. The discrepancy between the amount of calculated recharge at Ward Valley and other studies may be caused by differences in climatic regime, soil properties, or evaporation modeling techniques. In this case, unlike the previous scenario, the choice of soil properties and rainfall pattern are important to the results. In particular, the relationship between water content and matric potential (generally known as the water retention function) will strongly control the depth of wetting. If the soil texture is very coarse, water will infiltrate well below the zone of active evaporation and may easily become recharge. This dependence on texture makes the use of site specific soil data important in modeling studies involving absence of vegetation.

The rainfall pattern and intensity will also strongly control the magnitude of the recharge in this case. If rainfall events occur during periods of low evaporation, higher recharge can occur. **For this reason, the committee views the choice of the rainfall distribution used in the modeling to be less than conservative. Higher frequency and more intense rainfall scenarios may produce results significantly different from that reported.**

Scenario 7 and 7a: Complete Failure of the B/C Trench Cover

These scenarios were modeled to determine the transport possibility of radionuclides from the B/C trench through the unsaturated zone to an intruder well. The trench cover is assumed to fail completely and to allow infiltration of 88.1 cm of water into the waste from two hypothetical rainstorm and flood events. After this period of infiltration, no further infiltration is assumed and it appears that the cover mysteriously repairs itself. A background recharge rate of approximately 0.0003 cm/year is assumed throughout the unsaturated zone prior to infiltration. This rate is assumed from calculated values of the unsaturated conductivity at the water contents found in the upper 30 m of the profile. The assumptions in the modeling appear to be generally conservative with regard to the magnitude of the ponded infiltration and the mobility of the buried waste.

The results of the unsaturated-zone modeling were coupled with a solute transport model to determine the concentrations of various radionuclides reaching a well located at the site boundary. The analysis indicated that water and contaminants move quickly downward in initial response to the large infiltration rate. Following this phase of transport, both water and solutes diffuse slowly to the water table. Very low concentrations of long-lived, non-sorbing radionuclides were calculated to reach the water table within 10,000 years.

The results of the unsaturated-zone modeling depend strongly on the nature of infiltration at the land surface. In this scenario, infiltration of two large storms is followed by little or no infiltration for the remainder of the modeling period. Given that this scenario is based on a trench cap failure, the assumption of no infiltration following these two storms appears inappropriate. **The committee concludes that a much more credible scenario would include some enhanced infiltration following the cover failure, approaching the rate for an unvegetated surface. This would account for the failed cover and provide a credible "worst case" scenario for the facility.**

Conclusions Regarding the Adequacy of Modeling to Incorporate the Complexities of the System

1. **The assumption of homogeneity in the soil properties appears reasonable for performance modeling purposes, provided conservatism in all properties is maintained.** As additional data from the site and from other studies become available, these data should be incorporated and the performance modeling updated.

2. **The modeling for Scenario 1 is reasonable for the purpose intended. Additional modeling to include the effects of higher permeability and higher rainfall could be conducted, although it is unlikely to affect the results significantly.**

3. **The modeling results for Scenario 2 are in conceptual agreement with previous work showing that recharge can occur in arid climates if the vegetation is removed. The modeled recharge is lower than some studies and indicates the sensitivity of this modeling to actual field conditions. Given the uncertainties of the model assumptions and previous work, it is reasonable to assume that the model results represent a minimum rate of recharge and that actual values could be higher. This scenario should also be modeled to include higher permeability and rainfall intensity values, as they could be expected to have a significant effect.**

4. **The scenario of a failed cover on the B/C trench (Scenario 7 and 7a) is not conservative because of the very limited flux imposed from the surface following the failure of the cover. The committee concludes that the results of this modeling cannot be used to support the conclusion that transport through the unsaturated zone is negligible under all credible conditions.**

Summary of Modeling Subissue

The unsaturated zone beneath the Ward Valley site is complex and heterogeneous. The data collection efforts provide detailed data on the hydraulic properties in the upper 30 m of the unsaturated zone. **For modeling purposes, the assumption of homogeneity in the upper 30 m appears justified.**

The lack of data from below 30 m in the unsaturated zone makes it difficult to assess the impacts of the assumption of homogeneity deeper in the unsaturated zone, although it is unlikely to impact significantly the overall performance of the site. Such data, however, could provide significant assurance regarding the conservative nature of the performance modeling.

The performance modeling used to assess the failure of the B/C trench cover is not conservative because of the assumption of very limited infiltration following the failure. As this modeling is critical to the performance of the site at the compliance boundary, it is important that a conservative and credible infiltration amount be assigned to the failed cover.

SUMMARY OF CONCLUSIONS[3]

Major Conclusion

The committee concludes from multiple lines of evidence that recharge or potential transfer of contaminants through the unsaturated zone to the water table at the Ward Valley site, as proposed by the Wilshire group, is highly unlikely.

Basis for Committee Judgements and Conclusions

The committee reviewed multiple lines of evidence to evaluate subsurface water flux at the Ward Valley site. The committee based its conclusions concerning the unsaturated zone on the data, observations, and discussions in this chapter, including the following information.

- In 82 samples from near the surface to a depth of 27 m, water contents were generally very low (94 percent of the samples had water contents less than 10 percent, and 6 percent of the samples had water contents between 10 and 15 percent).
- Water content monitored in a neutron probe access tube installed to six meters depth showed that the maximum depth of penetration after rainfall was about one meter.
- Water potentials monitored by thermocouple psychrometers down to 30 m depth were very low (-3 to -6 Mpa).

---

[3]   Two committee members, J. Oberdorfer and M. Mifflin, dissented from this conclusion. Their statements can be found in Appendices E and F at the end of this report.

- Chloride concentrations measured in three boreholes to a depth of 30 m were very high (up to 15 g/l). The time required to accumulate these large quantities of chloride down to 30 m depth was calculated to be approximately 50,000 yr (Prudic, 1994b).
- Estimated water fluxes based on chloride data were very low (0.03 to 0.05 mm/yr below 10 m depth).

Measured tritium at 30 m depth, which could be interpreted as evidence for recent rapid downward migration of water, is not consistent with the foregoing soil-physics or chloride data, which indicate very dry conditions with very slow water movement and limited infiltration. **The committee concludes that the most likely explanation for the measured tritium is contamination with atmospheric tritium owing to inappropriate sampling procedures.**

In addition, the committee has reviewed data from similar regions and cited results of field experiments and related literature from other arid-region unsaturated zones to supplement the limited field evidence. We have resolved several inconsistencies in the data sets by attributing the source(s) of ambiguity to (a) inherent uncertainties in many of the measurements, (b) errors in procedural and collection methodologies, and (c) analysis and reporting errors.

Limitations of Field Data

The committee notes that monitoring hydraulic parameters in dry soils like those at the Ward Valley is very difficult, leading to several limitations in collecting field data. The restrictions imposed have three main causes: (1) the effects of low water fluxes; (2) the effects of instrument limitations in arid soils; and (3) the results of unresolved inconsistencies in the data and/or project decisions on where and how deep to test.

Effects of Low Water Fluxes

The soil physics and chloride data indicate that subsurface water fluxes in the upper 30 m of the unsaturated zone are extremely low. Collection of water for tritium analysis in these dry soils is quite difficult because of the large quantities of water required for the analysis relative to the very low water content of the soils. In addition, because of these very low water fluxes, it is difficult to resolve easily the rate and direction of water movement with available equipment and sampling procedures. **Based on the committee's experience and understanding of the unsaturated zone at Ward Valley, it is not currently possible to resolve definitely the exact magnitude and direction of the water flux. In the case of the Ward Valley site, qualitative terms such as very small and extremely slow can and should be used in place of more quantitative terms of water flux.**

Effects of Instrument Limitations

The committee also notes that monitoring hydraulic parameters in arid unsaturated zones is very difficult because of the lack of methods, procedures, and reliable instruments to measure precisely the hydraulic and hydrochemical parameters used to estimate water flux in dry desert soils. Some of the instruments, particularly the thermocouple psychrometers and heat dissipation probes, are not very robust and have a high failure rate. Although serious problems were encountered with instrumentation for water potential monitoring, all recorded water potential values were very low (-3 to -6 MPa) and are consistent with measured water contents. Matric potentials monitored by heat dissipation probes (-0.2 to -0.5 MPa) were much larger than the water potentials monitored by the thermocouple psychrometers (-3 to -6 MPa), indicating wetter conditions than would be expected from the water content values. **The committee attributes this inconsistency between the high matric potential values and low measured water contents to the wet silica flour used for installation of the heat dissipation probes.** In these dry soils, however, larger standard errors in the data can be tolerated, compared with much wetter soils, without significantly affecting the estimates of water flux because of the exponential decrease in hydraulic conductivity with decreased water content.

Quantity and Quality of Data

In part because of instrumentation and sampling problems, but also because of project decisions on where and how deep to test, the number and distribution of observations and the quantity of data collected for site characterization were very limited. **In the committee's view, more confirmatory information on spatial variability of hydraulic and hydrochemical attributes is highly desired to provide further assurance that the limited data from which site characteristics were determined are representative of the entire site both areally and vertically.**

In several instances, additional data and/or sampling will be required to interpret data correctly. This particularly applies to the tritium measurements in the unsaturated zone and to the apparent vertical gradient in the ground water beneath the site. These are inconsistent with most other data. Similarly, uncertainties in the measurements of soil-water potential and uncertainties generated by the limited hydraulic data from the unsaturated zone below 30 m can be reduced only through additional characterization of this zone. If the site is developed as a LLRW facility, resolution of these uncertainties will be important to the development of reliable baseline data for the planned monitoring during site operations. These issues, which relate to the monitoring program plan during site operations and beyond, are discussed in detail in Chapter 6 of this report.

Subissue Conclusions

With regard to the specific subissues raised (Wilshire et al., 1994):

1. **With respect to the measured soil properties of subissue 1, the committee concurs with the license application that water movement through the upper 30 m of the unsaturated zone is extremely slow. As explained previously, dry soils with low subsurface water fluxes, do not permit easy determination of the direction and rate of water movement; however, in the opinion of the committee, the most probable direction of water movement is upward, based on the low water potentials measured, as well as on what has been observed at other arid-zone sites with similar characteristics. The conclusions are also based on the observations of limited infiltration during the monitoring period, the low water contents and potentials found, and the significant accumulations of chloride in the upper 30 m of the unsaturated zone.**

   With respect to the modeling of the water movement in subissue 1, the committee determined that the performance modeling is consistent with the resolved data to encompass the expected range of variability and complexity of transport through the unsaturated zone, with the exception of the failure scenario for the B/C trench cover. For the B/C trench cover failure scenario, the committee has not determined if a more conservative scenario, that includes enhanced percolation through the cover, would result in significant risk at the compliance boundary.

2. **The committee concludes that site characterization has found no evidence for rapid deep migration of water along preferred pathways. Although it is impossible to demonstrate its absence conclusively, the lack of major surface features indicative of rapid flux, the significant quantities of soluble chloride found in several of the boreholes, low water content, low water potentials, the lack of evidence for recent recharge at the water table or for deep penetration of natural infiltration, do not support the hypothesis of rapid pathways.**

3. **The committee finds that the conclusion in the license application that gas diffusion is responsible for the tritium reported in the unsaturated zone is conceptually incorrect. The committee concludes that inappropriate sampling procedures most probably introduced atmospheric tritium into the samples. Except for three data points at depths of 5.1 m and 5.4 m, the tritium data are not distinguishable from zero owing to inadequate evaluation of the sample-collection blank. The three results from the uppermost sampling depths may represent atmospheric contamination, or they may indicate small amounts of shallow infiltration. Due to these uncertainties, the tritium data may not be used for evaluation of infiltration.**

4. **The committee concurs with both the DHS and the Wilshire group that significantly higher rates of percolation and recharge are possible beneath Homer Wash. Given its location downgradient from the site, the committee concludes that such potential recharge is unlikely to affect the trenches designed for the disposal facility.**

5. The committee concludes that the concentrations of isotopic tracers D, ^{18}O, and ^{14}C of ground water at Ward Valley do not support an interpretation of recent ground-water recharge by rapid movement of water through the unsaturated zone in the last 50-100 years, as proposed by the Wilshire group. The committee also concludes that the bulk ^{14}C age of the ground water exceeds 4,500 years.

RECOMMENDATIONS

In the committee's opinion, thick unsaturated alluvial sediments in arid environments such as are beneath the Ward Valley site are favorable hydrologic environments for the isolation of low-level radioactive waste. Both theoretical understanding and a growing body of field evidence demonstrate that the amount of water and rate of water movement throughout most of these unsaturated zones is very small and very slow, respectively. This does not imply, however, that siting of disposal facilities can be made without careful study and analysis, as it has also been found that percolation can occur in certain locations in arid regions, particularly where water is concentrated at the surface. Such concentrations occur in ephemeral washes and natural or artificial depressions. For this reason, site characterization data must be of sufficient quantity and quality to address the areal variability of percolation at a potential site to provide reasonable assurance that rapid movement of water through the unsaturated zone is unlikely.

The committee attributes some of the incomplete and/or unreliable data sets to the fact that hydrologic processes in arid regions are characterized by extreme events which do not follow a one-year calendar. For this reason, regulatory and/or budgetary guidelines that permit one-year or other short-duration time frames not suitable for arid-soil characterization can result in conflict between permissible regulatory timetables and the optimum time requirements for arid-region hydrologic investigations, which can easily lead to incomplete or ambiguous results. In the committee's opinion, characterization activities should receive priority over arbitrary regulatory timetables, or short time-frame budgetary constraints, particularly in arid regions.

General Recommendation

• To guard against deficiencies in characterization and monitoring efforts, and as more emphasis is placed on arid regions for waste disposal, the committee recommends, as a general rule, that an independent scientific peer review committee be established for such sites to provide oversight early in the permitting process, to help resolve scheduling conflicts, and to assess and suggest improvements in the site characterization plans and investigations. In this way, conflicts in, and other concerns with, characterization data from the unsaturated zone can be resolved as they arise. In

Chapter 6 of this report, this recommendation is also discussed with reference to monitoring.

   • The committee further recommends that such on-going peer review activities and recommendations be specifically included in the permitting applications to provide assurance to both the public and the scientific community of the expected performance of any waste-disposal site. Such independent review would also provide guidance to the regulatory community in assigning realistic and appropriate timetables for all phases of development. The committee hopes that other states and compacts consider these ideas and recommendations for future waste sites.

Specific Recommendations

   1. As both water-content and water-potential monitoring, and tritium analyses, are proposed for site and post-closure monitoring, the committee recommends that attempts be made to resolve or improve these data. To accomplish this, the following are recommended to establish base levels for monitoring:

   (a) additional sampling for tritium, water-content logging, and water-potential monitoring;

   (b) continued characterization of the unsaturated zone using corrected methodologies and equipment to provide a more complete data set for monitoring during site operation, and for an effective closure plan for the facility.

   (c) confirmatory data from a greater depth in the unsaturated zone below the present 30-m limit for confident monitoring of the site during operation.

   2. The committee recommends that to obtain more complete knowledge of the background levels of tritium in order to ensure that subsequent monitoring data can be adequately understood, and to develop action levels (see Chapter 6 for details), the following actions will be necessary:

   (a) because of the difficulty of determining tritium blank values for the air piezometer collection method, alternate collection techniques, such as vacuum distillation of core samples, should be considered;

   (b) sampling and analyzing for ^{36}Cl as a check on the tritium data. (The chemical tracer, ^{36}Cl, has been used successfully elsewhere to constrain the magnitude of liquid water movement in the unsaturated zone and should be considered. Sampling for ^{36}Cl can be accomplished much more readily than for tritium at the Ward Valley site because (1) large quantities of water are not required for analysis and (2) chloride concentrations in the soil water are extremely high.)

   3. The committee recommends that an analysis be conducted of a more conservative failure scenario for the B/C trench cover which includes enhanced percolation through the cover.

REFERENCES

Allison, G. B., and M. W. Hughes. 1983. The use of natural tracers as indicators of soil-water movement in a temperate semi-arid region. Journal of Hydrology 602:157-173.

Allison, G. B., G. W. Gee, and S. W. Tyler. 1994. Vadose-zone techniques for estimating groundwater recharge in arid and semi-arid regions. Soil Science Society of America Journal 58:6-14.

Baumgardner, R. W., Jr. and B. Scanlon. 1992. Surface fissures in the Hueco Bolson and adjacent basins, West Texas. University of Texas at Austin, Bureau of Economic Geology, Geological Circular 92(2):1-40.

Beven, K. 1991. Modeling preferential flow, *in* Preferential Flow, Proc. National Symposium, T.J. Gish and A. Shirmohammadi, eds. American Society of Agricultural Engineers. St. Joseph, Michigan. p. 77.

R. Briscoe. 1994. Personal Communication.

Butters, G. L., W. A. Jury, and F. F. Ernst. 1989. Field scale transport of bromide in an unsaturated soil 1. Experimental methodology and results. Water Resources Research 25(7):1575-1581.

California Department of Health Services. 1994. Response to Court Order dated June 24, 1994.

Campbell, G. S. 1985. Soil Physics with Basic: Transport Models for Soil-Plant Systems. New York: Elsevier. 150 pp.

Claussen, H. 1986. Late-Wisconsin paleohydrology of the west-central Amargosa Desert, Nevada, USA. Chemical Geology (Isotope Geoscience Section) 58:311-323.

Clayton, C. G. and D. B. Smith, 1963. A comparison of radioisotope methods for river flow measurement, *in* Radioisotopes in Hydrology. Proceedings of the Symposium on the Application of Radioisotopes in Hydrology. International Atomic Energy Agency, Tokyo, March 5-9. pp. 1.

Committee to Bridge the Gap. 1994. Presentation to the Committee to Review Specific Scientific and Technical Issues Related to the Siting of a Low Level Radioactive Waste Disposal Site in Needles, California. July 7.

Cook, P. G., B. R. Walker, and I. D. Jolly. 1989. Spatial variability in groundwater recharge in a semi-arid region. Journal of Hydrology 111:95-212.

Cook, P. G., W. M. Edmunds, and C. B. Gaye. 1992. Estimating paleorecharge and paleoclimate from unsaturated zone profile. Water Resources Research 28:2721-2731.

Craig, H. 1963. Isotopic variation in meteoric waters. Science 133:1702-1703.

de Marsily, G. 1986. Quantitative Hydrogeology. San Diego: Academic Press. 440 pp.

Dettenger, M. D. 1989. Reconnaissance estimates of natural recharge to desert basins in Nevada, USA by using chloride mass-balance calculation. Journal of Hydrology 106:55-78.

Estrella, R., S. Tyler, J. Chapman, and M. Miller. 1993. Area 5 site characterization project- report of hydraulic property analysis through August 1993. Desert Research Institute, Water Resources Center 45121: 1-51.

Fabryka-Martin, J., M. Caffee, G. Nimz, J. Southon, S. Wightman, W. Murphy, M. Wickman, and P. Sharma. 1993. Distribution of chlorine-36 in the unsaturated zone at Yucca Mountain: an indicator of fast transport paths. Focus '93 Conference. Site Characterization and Model Validation. American Nuclear Society. Las Vegas. pp. 56-68.

Fischer, J. M. 1992. Sediment properties and water movement through shallow unsaturated alluvium at an arid site for disposal of low-level radioactive waste near Beatty, Nye County, Nevada. U.S. Geological Survey, Water Resources Investigations Report 92(4032):1-48.

Fontes, J. C. 1992. Chemical and isotopic constraints on ^{14}C dating of ground water *in* Radiocarbon After Four Decades. R.E. Taylor, A. Long, and R.S. Kra, eds. New York: Springer-Verlag. 242-261

Freeze, R. A., and J. A. Cherry. 1979. Groundwater. New York: Prentice Hall. 604 pp.

Friedman, I., G. I. Smith, G. D. Gleason, A. Warden, and J. M. Harris. 1992. Stable isotope composition of waters in southeastern California: 1. Modern precipitation. Journal of Geophysical Research 97:5795-5812.

Fritz, P. and J. C. Fontes, editors. 1980. Handbook of Environmental Geochemistry. Volume 1. Elsevier. Amsterdam.

Gardner, W.R. 1958. Some steady-state solutions of the unsaturated moisture flow equation with application to evaporation from a water table. Soil Science, 85:228-232.

Gee, G. W., P. J. Wierenga, B. J. Andraski, M. H. Young, M. J. Fayer, and M. L. Rockhold. 1994. Variations in water balance and recharge potential at three western desert sites. Soil Science Society of America Journal 58:63-82.

Glass, R. J., J. Y. Parlange, and T. S. Steenhuis. 1991. Immiscible displacement in porous media: Stability analysis of three-dimensional, axisymmetric disturbances with application to gravity-driven wetting front instability. Water Resources Research 27(8):1947-1956.

Grant Environmental. 1994. Report prepared for U.S. Ecology entitled NETPATH analysis for the Ward Valley low-level radioactive waste project. Report dated September 27, 1994.

Gvirtzman, H. and M. Margaritz. 1986. Investigation of water movement in the unsaturated zone under irrigated area using environmental tracers. Water Resources Research, 22:635-642.

Harding Lawson Associates. 1990. Letter Report to U.S. Ecology entitled Evaluation of Tritium Transport Mechanisms. Dated December 20, 1990.

Harding Lawson Associates. 1994a. Letter Report to Ms. Ina Alterman; Interpretation of groundwater quality data; Ward Valley. California. Dated October 6, 1994.

Harding Lawson Associates. 1994b. Letter Report to Ms. Ina Alterman; Supplemental information regarding tritium vapor sampling. Dated October 12, 1994.

Harding Lawson Associates. 1994c. Letter Report to Ms. Ina Alterman; Summary of performance assessment modeling; Proposed low-level radioactive waste disposal facility. Ward Valley California. Dated October 12, 1994.

Helling, C. S., and T. J. Gish. 1991. Physical and chemical processes affecting preferential flow, *in* Preferential Flow, Proc. National Symposium, T.J. Gish and A. Shirmohammadi, eds. American Society of Agricultural Engineers. St. Joseph, MI. p. 77.

Hills, R. G., P. J. Wierenga, D. B. Hudson, and M. R. Kirkland. 1991. The second Las Cruces trench experiment: Experimental results and two dimensional flow predictions. Water Resources Research 27(10):2707-2718.

Hostettler, S. 1994. Presentation to the Committee to Review Specific Scientific and Technical Issues Related to the Siting of a Proposed Low Level Radioactive Waste Disposal Facility in Needles, California. July 7.

International Atomic Energy Agency. 1981. Statistical treatment of environmental isotope data in precipitation. International Atomic Energy Agency Technical Report Series 206. Vienna, Austria. 255 pp.

Junge, C. E. and R. T. Werby. 1957. The concentrations of chloride, sodium, potassium, calcium, and sulfate in rain water over the United States. Journal of Meteorology 15(3):417-425.

Kalin, R. M. 1994. The hydrogeochemical evolution of the groundwater of the Tucson Basin with applications to 3-dimensional groundwater flow modeling. Ph.D. dissertation. University of Arizona. 510 pp.

Lehmann, B. E., S. N. Davis, J. T. Fabryka-Martin. 1993. Atmospheric and subsurface sources of stable and radioactive nuclides used for groundwater dating. Water Resources Research. 89:2027-2040.

License Application. 1989. U.S. Ecology, Inc. Administrative Record, Ward Valley Low-Level Radioactive Waste Disposal Facility, Sections 2420.B and 2600.A; Appendices 2420.B, 2500.A.

Long, A. and R. M. Kalin. 1990. A suggested quality assurance protocol for radiocarbon dating laboratories. Radiocarbon, 32(3):329-334.

Mazor, E. 1991. Applied Chemical and Isotopic Groundwater Hydrology. Open University Press. Buckingham, UK. 274 pp.

Münnich, K. O. and W. Roether, 1963. A comparison of carbon-14 and tritium ages of groundwater, *in* Radioisotopes in Hydrology. Proceedings of the Symposium on the Application of Radioisotopes in Hydrology. International Atomic Energy Agency, Tokyo, March 5-9. pp. 397.

Nichols, W. D. 1987. Geohydrology of the unsaturated zone at the burial site for low-level radioactive waste near Beatty, Nye County, Nevada. U.S. Geological Survey, Water Supply Paper 2312, pp 57.

Östlund, H. G., 1989. Written Communication. November 1.

Phillips, F. M. 1994. Environmental tracers for water movement in desert soils of the American southwest. Soil Science Society of America Journal 58:15-24.

Phillips, F. M., J. L. Mattick, T. A. Duval, D. Elmore, and P. W. Kubick. 1988. Chlorine-36 and tritium from nuclear-weapons fallout as tracer for long-term liquid and vapor movement in desert soils. Water Res. Research 24:1877-1891.

Plummer, L. Niel, Eric C. Prestemon and David L. Parkhurst. 1991. An Interactive Code (NETPATH) for Modeling NET Geochemical Reactions along a FlowPATH. U.S. Geological Survey Water Resources Investigations Report 91-4078. 101 pp.

Prudic, D. E. 1994a. Effects of temperature on water movement at the arid disposal site for low-level radioactive wastes near Beatty, Nevada. Geological Society of America, Abstracts with Programs, 26:A-391.

Prudic, D. E. 1994b. Estimates of percolation rates and ages of water in unsaturated sediments at the Mojave Desert Sites, California-Nevada, U.S. Geological Survey Water Resources Investigations Report 94-4160. 19 pp.

Rozanski, K. L., L. Araguás-Araguás, and R. Gonfiantini. 1992. Relation between long-term trends of oxygen-18 isotope composition of precipitation and climate. Science. 258:981-985.

Scanlon, B. R. 1992a. Moisture and solute flux along preferred pathways characterized by fissured sediments in desert soils. Journal of Contaminant Hydrology 10:19-46.

Scanlon, B. R. 1992b. Evaluation of liquid and vapor water flow in desert soils based on chlorine 36 and tritium tracers and nonisothermal flow simulations. Water Resources Research 28:285-297.

Scanlon, B. R. 1994. Water and heat fluxes in desert soils, 1. Field studies, Water Resources Research 30:709-719.

Scanlon, B. R. and P. C. D. Milly. 1994. Water and heat fluxes in desert soils, 2. Numerical simulations. Water Resources Research 30(3):721-733.

Smith, G. I., I. Friedman, G. D. Gleason, and A. Warden. 1992. Stable isotope composition of waters in southeastern California: 2. Ground waters and their relation to modern precipitation. Journal of Geophysical Research 97:5813-5823.

Sophocleous, M. and C. A. Perry. 1984. Experimental studies in natural groundwater recharge dynamics: Assessment of recent advances in instrumentation. Journal of Hydrology 70:369-382.

Steenhuis, T. S., J. Yves Parlange, and J. A. Aburime. 1994. Preferential flow in structured and sandy soils: Consequences for modeling and monitoring, *in* Handbook of Vadose Zone Characterization and Monitoring, Wilson, L. G., L. G. Everett, S. J. Cullen, eds. Lewis Publishers, Boca Raton. pp. 61-77.

Sully, M. J., T. E. Detty, D. O. Blout, and D. Hammermeister. 1994. Water fluxes in a thick desert vadose zone. Geological Society of America Annual Meeting, October 1994. Abstracts with Program, 26:A-391.

Taylor, J.K. 1987. Quality Assurance of Chemical Measurements. Lewis Publishers, Inc. Chelsea, Michigan. 328 pp.

Turton, D. J., D. R. Barnes, and J. Navar. 1995. Old and new water in subsurface flow from a forest soil block. Journal of Environmental Quality. 24:139-146.

Tyler, S. W. and G. R. Walker. 1994. Root zone effects on tracer mitigation in arid zones. Soil Science Society of America Journal 58:25-31.

U.S. Ecology. 1990. Supplemental Unsaturated Zone Data Report, Administrative record of the Ward Valley Proposed Low Level Radioactive Waste Disposal Site.

Vogel J. C. and D. Ehhalt. 1963. The use of the carbon isotopes in groundwater studies, *in* Radioisotopes in Hydrology. Proceedings of the Symposium on the Application of Radioisotopes in Hydrology. International Atomic Energy Agency, Tokyo, March 5-9. pp. 383.

Vogel J. C., D. Ehhalt, and W. Roether. 1963. A survey of the natural isotopes of water in South Africa, *in* Radioisotopes in Hydrology. Proceedings of the Symposium on the Application of Radioisotopes in Hydrology. International Atomic Energy Agency, Tokyo, March 5-9. pp. 407.

Weeks, E. P., D. E. Earp, and G. M. Thompson. 1982. Use of atmospheric fluorocarbons F-11 and F-12 to determine the diffusion parameters of the unsaturated zone in the southern high plains of Texas. Water Resources Research 18(5):1365-1378.

Wierenga, P. J., J. M. H. Hendricks, M. H. Nash, J. Ludwig, and L. Daugherty. 1987. Variation of soil and vegetation with distance along a transect in the Chihuahuan Desert. Journal of Arid Zone Environments, 13(1):53-64.

Wierenga, P. J., R. G. Hills, and D. B. Hudson. 1991. The Las Cruces Trench Site: Characterization, experimental results, and one-dimensional flow predictions. Water Resources Research 27:2695-2705.

Wilshire, H. G., K. A. Howard, and D. M. Miller. 1993a. Memorandum to Secretary Bruce Babbitt, dated June 2.

_____1993b. Description of earth science concerns regarding the Ward Valley low-level radioactive waste site plan and evaluation. Released December 8.

Wilshire, H. G., K. A. Howard, D. M. Miller, K. Berry, W. Bianchi, D. Cehrs, I. Friedman, D. Huntley, M. Liggett, and G. I. Smith. 1994. Ward Valley Proposed Low-Level Radioactive Waste Site: A Report to the National Academy of Sciences. Presentations made to the review committee on July 7-9 and August 30-September 1.

Yang, I. C. 1992. Flow and transport through unsaturated rocks - data from two test holes, Yucca Mountain, Nevada. High Level Radioactive Waste Management: Proceedings of the Third International Conference. American Nuclear Society. pp. 732-737.

Young, M. H., P. J. Wierenga, R. G. Hills, J. Vinson. 1992. Evidence for piston flow in two large scale field experiments, Las Cruces, New Mexico. EOS, Trans. AGU 73:156.

4

INFILTRATION AND LATERAL FLOW

Issue 2
The Potential Infiltration of the Repository Trenches
by Shallow Subsurface (Lateral) Flow

THE WILSHIRE GROUP POSITION

According to Wilshire et al. (1994) the potential for shallow lateral flow of water downslope into the waste trenches and from the trenches was not addressed in any of the site evaluation documents. Specifically, the Wilshire group suggested that available data show that shallow, low-permeability layers may exist in the alluvial fan slope beneath the site and toward the main valley drainage at Homer Wash. In their view, these could promote lateral flow, leading to excess water leaking into the trenches and migration of contaminants from the trenches to Homer Wash. Once in Homer Wash, these contaminants could be redistributed into the general environment by wind and water erosion much faster than by percolation to the water table.

THE DHS/U.S. ECOLOGY POSITION

The position of the California Department of Health Services (Brandt, 1994) is that "there is little, if any, deep percolation of water through the thick, highly moisture-deficient sediments in the vadose zone underlying the Ward Valley site." According to DHS, 'the infiltrating precipitation is either removed by evapotranspiration or held in storage in the highly moisture-deficient soils near the ground surface until such time as it is removed by evapotranspiration." DHS further stated that their findings showed "the caliche layers at the site to be discontinuous and permeable such that one would not expect them to perch infiltration water in undisturbed areas where they are overlain by sediments. However, where exposed on the surface in the abandoned highway borrow pit, the shallow caliche layers act as a form of discontinuous pavement and the resulting rapid runoff of surface water has led to ponding in the low end of the borrow pit" (Brandt, 1994).

THE COMMITTEE'S APPROACH

The committee reviewed the data and information pertinent to the issue to evaluate subsurface lateral flow in natural desert soils under natural and enhanced rainfall conditions, ponded conditions, and in engineered systems. Data from Ward Valley and other pertinent literature, including results of field studies on subsurface flow at sites other than Ward Valley, were used by the committee to clarify the issue.

Two basic scenarios are examined: lateral flow under rainfall conditions and lateral flow under ponded conditions. Under rainfall conditions, large areas of soil would be wetted, but the slope of the calcic soil horizon would be an important factor in controlling the extent of lateral flow. Although lateral flow is more likely under ponded conditions because of the higher fluxes, the localized nature of the ponded conditions, the amount of water ponded, and the lithologic continuity of layers are important in determining the extent of lateral flow.

LATERAL FLOW UNDER NATURAL AND ENHANCED RAINFALL CONDITIONS AT ARID SITES

Lateral flow under natural conditions in arid soils depends on many factors, including (1) the lateral continuity of a perching (or low-permeability) horizon, (2) the relative permeabilities of the soil horizons at the site, (3) the magnitude of the rainfall events, (4) the storage capacity of the soil overlying the less permeable layer, and (5) the slope of the less permeable layer.

1. Lateral continuity: The development of a low-permeability soil layer depends on the long-term stability of the geomorphic surface. Roy J. Shlemon's soil report (LA Appendix 2310.A) states that the site was typified "by geomorphic stability throughout late Quaternary time." He states that the age of the surface that is reflected in the soils is at least 35,000 to 40,000 years. This stability would allow the development of a laterally continuous unit.

The trench logs are ambiguous. They mention "pervasive" carbonate and show a laterally continuous white carbonate layer (North Wall of WV-TP-1), while the South Wall was not excavated to sufficient depth (>0.6 m) to reach the depth of that horizon. The trench log for WV-TP-2T shows a discontinuous white carbonate-cemented horizon at the same depth, although a slightly deeper horizon (Qf2v) is described as containing pervasive carbonate cement. A still deeper (approximately 2.4 m depth) paleosol is described in Shlemon's soil log as "very hard" and "violently effervescent." Both trenches were located downslope from the proposed waste repository site. Brief field observations of the borrow pit located upslope of the site indicate that the calcrete[1] is a continuous horizon. It also appears to be continuous on aerial photographs, at least in the western portion of the pit up to the "bench" east of which it has been excavated. It is possible (more supported than contradicted by available information) that the horizons are laterally continuous.

2. Relative permeability of soil horizons: For ponding to occur, the calcrete layer must be less permeable than the soil above so that it acts as a (leaky) impedance layer which could temporarily perch under high infiltration rates during prolonged rainfall. Two observations in the borrow pit suggest that it is less permeable. First, the fact that the calcrete

---

[1] See Box 2.2 in Chapter 2 of this report for a discussion of soil carbonates.

layer can sustain fractures over seven feet long indicates that it is highly cemented. The second is that evidence of ponding is present at both the western and eastern ends of the borrow pit (Harding Lawson Associates, 1994). At least under heavy rainfall conditions the calcrete layer, when exposed to the surface, is sufficiently impermeable to create temporarily standing water. It is not known, however, for how long bare calcrete horizons can sustain standing water, as calcrete layers can be slightly permeable.

Gile (1961) measured infiltration rates with a ring infiltrometer directly on calcrete layers classified as weak, moderate, strong, and very strong. Carbonate contents ranged from less than 1 to 93 percent. Infiltration rates of the calcic horizons ranged from 0.13 cm/hour to 15.0 cm/hour. Lowest infiltration rates were obtained where laminae occur in the uppermost part of the calcic horizon.

Further evidence for water movement through a dense calcrete horizon under natural conditions was provided by information on the depth distribution of bomb-pulse ^{36}Cl at the Nevada Test Site (Gifford, 1987). Some of the bomb-pulse tracer was found within the dense calcrete, which indicates water movement into the calcrete during the past 30 years. Experiments at the Las Cruces trench site have also demonstrated that calcrete layers are not impermeable and that significant vertical water flow through calcrete layers is possible (Wierenga et al., 1991). These field experiments showed that only limited lateral water movement occurred under enhanced rainfall conditions in a layered soil system that contained calcic horizons.

In addition to the possibility of soil layering and resultant changes in permeability causing lateral flow, another process called tension-dependent anisotropy can also promote lateral flow. Yeh et al. (1985) used a stochastic analysis of the flow equation to demonstrate that the ratio of horizontal to vertical permeability depends on the wetness of the soil profile, increasing as the wetness decreases or as the tension increases. Such an increase in the permeability ratio should promote lateral flow.

The Las Cruces experiments were designed to test, under unsaturated-flow conditions, the extent of lateral flow resulting from tension-dependent anisotropy and also from layering. The field experiments, however, showed limited lateral flow. The tests were conducted at the Las Cruces test site in the Jornada Range in southern New Mexico. Three comprehensive tests were conducted on this well-instrumented and characterized site. The results of the first experiment showed water fronts moving mostly downward with time after artificially being rained upon at a rate of 1.8 cm/day for 86 days (Figure 4.1). The area being rained upon was 4 m wide by 9 m long. A trench was dug to 6 m depth perpendicular to the long axis of the trench to observe and measure the water front penetrating the soil. The data in Figure 4.1 show that lateral flow at this site is limited. After 50 days and 91 cm of water application (7 times the total annual rainfall at Ward Valley), the water front had reached the 5.3 m depth. Lateral spreading after 63 days was only 1 m outside the rained-upon area. The extent of lateral water movement was also limited during the second experiment, which used an artificial rainfall rate of 0.5 cm/d. The third experiment again used an artificial rainfall rate of 1.8 cm/d for 86 days, similar to the first experiment. A total of 155 cm (61 in) of water was applied. Even though the soil was highly heterogeneous with several calcrete layers, three buried soil horizons, and a surface slope of 4 percent, lateral spreading was limited to 3 m

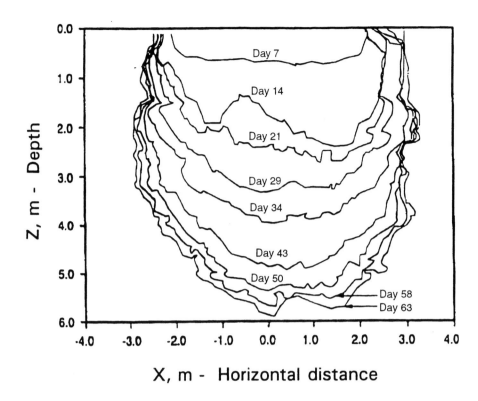

Figure 4.1 Advance of the wetting front with time following wetting of a 4 m wide experimental plot at the Las Cruces Trench site. The plot was rained upon for 86 days at a rate of 1.82 cm/day. Solid lines show distribution of water front on trench face from 7 to 63 days after initiation of rain event (Wierenga et al., 1991).

beyond the rained-upon area after 525 days (Vinson et al., 1995). Furthermore, saturation was not observed anywhere in the soil profile during the three experiments, and no water was observed to seep from any of the four trench faces.

    3. Magnitude of rainfall event: Under normal rainfall conditions, it is unlikely that sufficient water would infiltrate to create saturated flow. Computer simulations with 5 inches of rain falling over a 2-hour period in a soil with similar hydraulic characteristics as the Ward Valley site and under a slope of 2 percent has shown no saturated conditions over the less permeable layer located at 50 cm below the soil surface (Pan and Wierenga, 1995). Only when a 5-inch rainstorm was followed by another 5-inch rain fall the following day was saturation observed for less than 12 hours above the less permeable calcrete layer. Two consecutive 5-inch rainstorms within a 24-hour period are highly unlikely.

    4. Storage capacity of the soil: The storage capacity of soils was discussed earlier in Chapter 3 of this report. A surface layer at least 50 cm thick can store enough water to prevent ponding above the less permeable layer. Highly unlikely rainfall conditions such as simulated above, however, could induce subsurface ponding. Only in case of a very thin

surface layer or no surface layer at all, is ponding likely to occur under natural climate conditions. The combined effects of relatively permeable surface soil with a somewhat permeable calcrete layer, dry initial conditions of the surface layer, and the lack of extreme rainfall events (e.g. two 5-inch storms in a 24-hour period) make saturation above the calcrete layer unlikely.

5. Slope of the less permeable layer: Slopes along alluvial fans, such as at Ward Valley, could possibly promote lateral flow. Hillslope lateral flow may be observed in road cuts where water seeps out of soil layers. This is often caused by rainwater percolating into soil upslope, reaching a low-permeability layer, and nearly saturating the soil above the low-permeability layer. If the low-permeability layer has a significant slope (e.g., 10-30 percent), water will move downslope above the layer. This downslope flow is the immediate consequence of the downslope component of gravity. As explained by Philip (1993), "when you have a car in neutral with the safety brake off, it rolls down hill." However, in flat lands, defined by Miyazaki (1993) as land having a slope of less than 3 percent, flow is driven by capillary diffusion, and is predominantly downward (Figure 4.1).

At Ward Valley, the surface slopes about 2 percent to the east. Near-surface units are likely to have similar slopes, and sub-soil horizons should closely parallel the exposed surface. Drilling and test pits reveal local heterogeneity which makes it difficult to establish broad patterns of slope in the near-surface sediments. However, observations within about 1 to 2 meters of the surface in many areas of the fan indicate that the overall slope of the calcretes is not much different from the present slope and close to 2 percent. Such small slopes are too low (~ 2 percent) for the process of lateral flow in the upper unsaturated zone to be significant. Water will move down through the variously textured layers, rather than laterally.

In order to constrain the maximum flow rate that could be expected into the trench by lateral flow, let us assume that a completely impermeable layer is located at some shallow depth below the land surface. Overlying this impermeable layer is a layer of soil with hydraulic conductivity similar to that found at the Ward Valley site. From the infiltration test conducted during the characterization efforts, we can assume that the near surface soils have a saturated conductivity of approximately 1 m/day. We then take this layered soil system to be inclined such that the slope of the impermeable layer is similar to that found at the proposed site (2 percent) and intersected by a trench wall.

If we ignore the effects of capillarity, a simple estimate of the amount of water which can flow along the top of the permeable layer can be calculated from Darcy's law (Equation 3.1) as:

$$Q = K_{sat}AdH/dz$$

where Q is the volumetric flow, K_{sat} is the saturated conductivity of the soil, A is the cross sectional area of the flow, and dH/dz is the hydraulic gradient.

If we assume that the trench face is 100 m wide and that the depth of saturation above the impermeable layer is 5 cm (corresponding to approximately 2.5 cm of precipitation being ponded on the layer), the cross section available for flow is 5 m^2. Since the flow is ponded on the layer, and ignoring capillary forces, we can assume that the gradient is equal to the slope of the layer or 0.02 (2 percent). The resulting volumetric rate into the trench is therefore (1 m/day) (5 m^2) (0.02) = 0.1 m^3/day or approximately 100 liters per day. The actual rate would probably be much less as capillary forces would tend to hold much of the water within the soil matrix.

If such water were to flow into the trench continuously, an additional 36,500 liters of water would be available each year to contact the waste and potentially move downwards towards the water table. Assuming this water were spread over the entire area of the trench floor (approximately 400 m long by 400 m wide or 40,000 m^2), the resulting recharge rate through the trench floor would be 36 m^3/yr/40,000 m^2 or approximately 1 mm/yr. While this represents a considerable percentage increase in the recharge rate over what is proposed in the license application, the rate is still extremely small and is unlikely to affect the overall ability of the unsaturated zone to isolate waste.

The above model and discussion represent a simplified model of lateral flow. We have ignored the effects of capillarity and assumed that ponding occurs on a shallow, continuous, and completely impeding layer. Under the more realistic field conditions, capillarity and finite permeability of any impeding layer will significantly reduce the amount of water available for lateral flow. As a result, ponding above a caliche layer over an extended time period and significant lateral flow into the trench is highly unlikely. **Therefore, in the committee's opinion, hillslope flow is not significant at the Ward Valley site because the slopes are too low (~ 2 percent) to cause large amounts of lateral flow. In addition, as discussed in Chapter 3 of this report, water content in the unsaturated zone at the site is low. Under low water-content conditions, the very small downslope gravity component of subsurface flow is negligible compared to the diffusion component.**

LATERAL FLOW UNDER PONDED CONDITIONS AT ARID SITES

Although the natural interstream setting at Ward Valley shows no evidence for ponded conditions, information from ponding experiments provides conservative estimates of the degree of lateral flow if ponding occurs.

The infiltration test conducted at Ward Valley consisted of ponding water in a 4.5 m by 4.5 m infiltration pond. The pond was located in an excavation constructed to a depth of 1.8 m below ground surface (Ohland and Lappala, 1990). A total of 97 cm depth of water was added, most of which infiltrated in approximately 25 hours. Thus, while the water at the Las Cruces site was added relatively slowly, water at Ward Valley was ponded and allowed to infiltrate at maximum rate. After 139 days the water front had reached a depth of about 5 m below the surface of the pond. The initial soil-water content at the Ward Valley site was 6 percent by volume. Wetting of the soil, following ponding, was observed 1 m from the pond boundary, but no wetting was observed at 5 m from the edge of the pond. The limited extent of lateral water movement is attributed to the dominant effect of gravity and to variation in hydraulic conductivity with water content. While a large matric potential gradient appears at the margin of the wetted region, the exponential decrease in hydraulic conductivity with a reduction in matric potential results in low lateral fluxes from the wetted region.

A detailed ponding experiment was conducted at Yucca Mountain to evaluate subsurface flow in a layered desert soil system (Guertal et al., 1994). The experiment consisted of ponding 10 cm of water within a 3.5 m diameter ring for 14 days. Water content was monitored with a neutron probe installed in the center of the pond. Neutron probe access tubes were not installed to monitor lateral flow. Three distinct calcrete horizons were identified in the upper 8 m section. Subsurface water movement was restricted at each calcrete horizon. The downward flux was greater than the permeability of the calcrete, and water moved laterally on these low permeability horizons until the permeability of the calcrete could accommodate the downward flux. The data showed that it took only three hours to penetrate the top one meter thick calcrete. Results of this experiment indicate that calcrete will, upon wetting, begin to accept water, thereby limiting lateral flow. The experiment also demonstrates that the calcrete horizons are not impermeable.

LATERAL FLOW IN ENGINEERED SYSTEMS AT ARID SITES

In many instances trench caps are specifically designed to promote lateral flow. Clay layers or native soil amended with bentonite, as proposed for the Ward Valley B/C trench covers, are used to minimize downward water movement. Clay soil barriers are generally wetted and compacted to achieve permeabilities of approximately 10^{-7} cm/s. In arid settings, however, this wetted clay layer may become ineffective with time because it will dry and crack, creating potential preferred pathways for flow. This could enhance rather than minimize problems associated with downward water flow or upward radon diffusion because it would concentrate movement in the cracks.

Many trench covers, including the proposed B/C trench cover at Ward Valley, are designed with capillary barriers. The capillary barrier consists of coarse-grained sediments overlain by fine-grained sediments (Ross, 1990). The controlling mechanism is the capillary potential of the fine-grained soil, which prevents water from entering the larger pores of the underlying coarse-grained soil. At equilibrium, the matric potential of the two layers will be equal, and if the soils are sufficiently dry, the hydraulic conductivity of the coarse layer will be much less than that of the overlying fine layer, which will cause water to accumulate at the interface of the two layers. In humid sites, steep slopes (20 percent) are used to promote lateral water movement to a drainage collection system. Although the details vary, the concept of capillary barriers has proved valid through field studies and numerical modeling (O'Donnell and Lambert, 1989; Nyhan et al., 1993; IT Corporation, 1994). In arid sites, because of low fluxes, capillary barriers can be used to retard subsurface flow and allow greater time for water to evapotranspire. The slopes used at some of these sites are on the order of a few percent (Cadwell et al., 1993). However, by changing the composition of the fine layer over the lower layer, so that the horizontal conductivity is much larger than the vertical conductivity, lateral flow could be enhanced, even for gentle slopes (<5 percent) (Stormont, 1995).

Capillary barriers also exist in the natural system. The concept of capillary barriers explains why buried gravel paleochannels, which could constitute zones of preferred lateral flow under saturated flow conditions, do not act as preferred pathways in unsaturated systems. This is because, in unsaturated systems, the permeability of coarse sediments is less than that of fine sediments and gravel paleochannels retard downward flow, except near saturation.

The disposal facility at Ward Valley is designed so that the trench floor is sloped and any water that accumulates will be removed. If water ponded on the trench floor and was not removed, the infiltration experiments described above suggest that this water would move predominantly in a vertical direction. However, if the trench were to intersect a horizontal coarse-grained layer and became ponded, lateral flow would occur along this coarse layer. The extent of such lateral flow would be limited by the horizontal continuity of buried alluvial fan systems that produce these conductive zones. Even if the unit were laterally continuous, the extent of lateral flow would be controlled by the volume of ponded water, and by the hydraulic conductivities of the coarse and underlying fine-grained layer.

CONCLUSIONS

1. **Low water contents, low water potentials, and high chloride concentrations indicate that subsurface water fluxes are negligible in interstream settings under natural conditions in arid regions. The permeability of calcrete is high enough that such horizons probably do not prevent downward movement of water at these low fluxes. This is supported by chemical tracer data at the Nevada Test Site where bomb-pulse ^{36}Cl is found within a calcrete horizon. Experiments conducted at Las Cruces show that even under conditions of artificially enhanced rainfall (7 times the annual rainfall at Ward Valley) the permeability of calcrete horizons was sufficient to allow water to move predominantly downward and lateral flow was limited.**

2. **The committee concludes that the slope at the Ward Valley site at ~ 2 percent is too low to produce significant volumes of lateral flow in the unsaturated zone under natural conditions. The horizontal component of the gravity vector is simply too small on a 2 percent slope to overcome the resistance to flow through soils found at the site and to cause significant lateral flow as compared to downward flow that responds to the much larger vertical component of gravity.**

3. **Under localized ponding conditions, lateral water movement was limited during the ponding experiment at Ward Valley. This is attributed to the dominance of gravitational flow and to the exponential decrease in hydraulic conductivity laterally from the wetted zone. A ponding experiment conducted at Yucca Mountain showed that calcrete horizons restricted downward flow, although the data showed that in only 3 hours water penetrated the upper 1 m thick calcrete.**

In summary, the committee concludes that the moisture-deficient nature of the soils at Ward Valley, the absence of an effective slope factor, the very low subsurface water fluxes, and the limited ability of the calcareous horizons to impede vertical flow, eliminate lateral flow as a significant issue at the Ward Valley site. The committee emphasizes, however, that conditions that could cause local lateral flow, such as ponding and enhanced percolation through the runoff-control structure, should be avoided, particularly in and immediately surrounding the trenches.

REFERENCES

Brandt, E. C. 1994. Summary submittal of the California Department of Health Service. October 6, 1994. pp. 43.

Cadwell, L. L., S. O. Link, and G. W. Gee. 1993. Hanford Site Permanent Isolation Surface Barrier Development Program. Fiscal Year 1992 and 1993 Highlights, Report No. PNL-8741/UC-702, Pacific Northwest Laboratory, Richland, Washington.

Gifford, S. K., III. 1987. Use of chloride and chlorine isotopes to characterize recharge at the Nevada Test Site. Masters thesis, University of Arizona. 73 pp.

Gile, L. H. 1961. A classification of Ca-horizons in soils of a desert region, Dona Ana County, New Mexico. Soil Science Society of America Proceedings. 25:52-61.

Guertal, W. R., A. L. Flint, L. L. Hoffman, and D. B. Hudson. 1994. Characterization of a desert soil sequence at Yucca Mountain, NV. Proceedings, International High Level Nuclear Waste Conference, Las Vegas, NV. May 22-26, 1994. American Nuclear Society, LaGrange Park,IL pp.2756-2765.

Harding Lawson Associates. 1994. Summary of Borrow Area Investigation, Low Level Radioactive Waste Disposal Facility, Ward Valley, California. Report to Ina Alterman, NRC. Dated September 30.

IT Corporation. 1994. Use of engineered soils and other site modifications for low-level radioactive waste disposal. National Low-Level Waste Management Program. U.S. Dept. of Energy. DOE/LLW-207.

License Application. 1989. U.S. Ecology, Inc. Administrative Record, Ward Valley Low-Level Radioactive Waste Disposal Facility, Section 2310.A

Miyazaki, T. 1993. Water flow in soils. Marcel Dekker Inc., NY, 296 pp.

Nyhan, J. W., G. J. Langhorst, C. E. Martin, J. L. Martinez, and T. G. Schofield. 1993. Field studies of engineered barriers for closure of low-level radioactive waste landfills at Los Alamos, New Mexico, USA. Proc. of the 1993 International Conference on Nuclear Waste Management and Environmental Remediation, Prague, Czechoslovakia.

O'Donnell, E. and J. Lambert. 1989. Low-level radioactive waste research program plan. NUREG-1380. Washington, DC.

Ohland, G. L. and E. G. Lappala. 1990. Supplemental ground water flow and transfer mechanisms report. HLA Job. #171G0,068.11. Appendix 6151 B. U.S. Ecology, Inc.

Pan, L. and P. J. Wierenga. 1995. Unpublished modeling study for this report.

Philip, J. R. 1993. Comment on "Hillslope infiltration and lateral downslope unsaturated flow" by C.R. Jackson. Water Resources Research 29:4167.

Ross, B. 1990. The diversion capacity of capillary barriers. Water Resources Research. 26:2625-2629.

Stormont, J. C. 1995. The effect of constant anisotropy on capillary barrier performance. Water Resources Research. 31:783-785.

Wierenga, P. J., R. G. Hills, and D. B. Hudson. 1991. The Las Cruces Trench Site: Characterization, experimental results, and one-dimensional flow predictions. Water Resources Research 27:2695-2705.

Wilshire, H. G., K. A. Howard, and D. M. Miller. 1994. Ward Valley proposed low level radioactive waste site. A report to the National Academy of Sciences.

Vinson, J., P. J. Wierenga, R. G. Hills, and M. H. Young. 1995. Flow and transport at the Las Cruces Trench Site: Experiments IIb. University of Arizona, Department of Soil and Water Sciences Contract Report. 216 pp.

Yeh, T. C. J., L. W. Gelhar, and A. L. Gutjahr. 1985. Stochastic analysis of unsaturated flow in heterogeneous soils. 1. Statistically isotropic media. Water Resources Research 4:447-456.

5

GROUND-WATER PATHWAYS TO THE COLORADO RIVER

Issue 3
The Potential for Hydrologic Connection
Between the Site and the Colorado River

THE WILSHIRE GROUP POSITION

The Wilshire group's concern with a hydrologic connection between Ward Valley and the Colorado River arises from their concern that water may carry radionuclides from the waste down to the ground water. The Wilshire group identified a number of possible routes by which the ground water at Ward Valley could reach the Colorado River. As the water table beneath Ward Valley is 300 m higher in elevation than the Colorado River, the Wilshire group considers it possible that ground water flows from Ward Valley to other ground-water basins that discharge directly into the Colorado River. On the basis of interpretations of surface geology, subsurface materials, and data from wells and springs, they generalized these routes into five possible pathways (Figure 5.1). It is their opinion that "the topography of the water table must be established before anyone can establish whether these pathways or others exist and estimates of ground-water flow rates must be established to evaluate any potential hazard to the Colorado River from contaminated ground water" (Wilshire et al., 1994). They argued that additional data are needed before these routes could be eliminated as potential pathways.

In the absence of information on the elevation of the water table between the site and the Colorado River, the Wilshire group based their analyses on the identification of the major hydrogeologic units, estimation of the probable permeability characteristics of each unit, and the determination of the subsurface extent of each unit. They made use of gravity data to infer the subsurface distribution of their hydrogeologic units. They assumed that: (1) Quaternary and Tertiary alluvial sediments have a high permeability, (2) Tertiary volcanic rocks have somewhat lower and more variable permeability, but locally have significant fracture permeability, (3) gneiss and granite above the detachment fault system locally have significant fracture permeability, and (4) footwall rocks beneath the detachment surface are relatively impermeable.

The Wilshire group assessed the likelihood of these routes as follows (Wilshire et al., 1994):

Even if all bedrock is impermeable, ground-water routes south through Ward Valley and through low-density materials in low-relief divides to Rice (route 1) and Vidal Valleys (route 2) still are probable. If fractured high-density rocks above the detachment fault are permeable, routes east or southeast to Chemehuevi Valley and Piute Valley (routes 3 and 5)

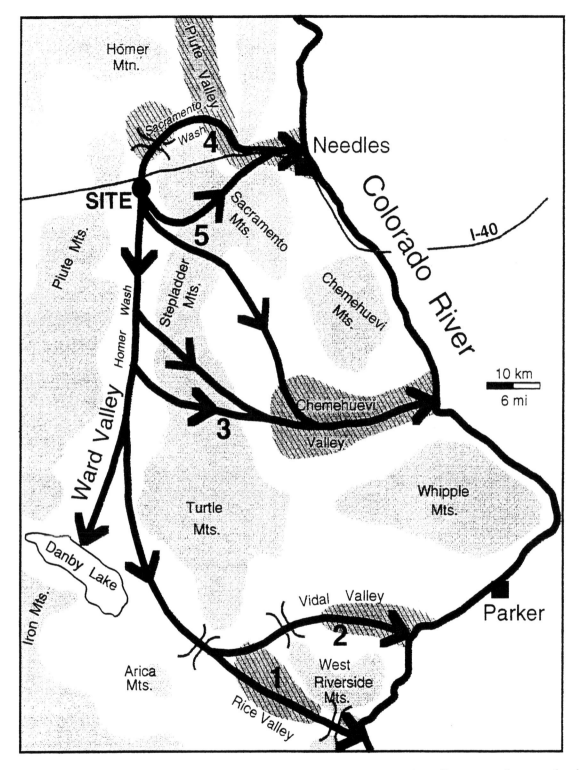

Figure 5.1 Postulated ground-water pathways from the Ward Valley ground-water basin to the Colorado River (Wilshire et al., 1994). Gray indicates mountains; cross-hatching denotes valleys.

are viable. If drawdown by pumping takes place in basins up-gradient from the site, the hydrologic gradient could be reversed so ground water at the site could communicate with that leading to or under Piute Valley (route 4).

Details of Pathways

Details of the Wilshire group's viewpoints on each of these routes can be summarized as follows:

Route 1: Ground-water flow from Ward Valley may bypass Danby Dry Lake and enter Rice Valley (Figure 5.1). The entire route from the site to the Colorado River could be within basin-fill deposits. No bedrock projects high enough to interfere with a continuous slope of the water table from Ward Valley to the Colorado River. The only possible interference from bedrock is the divide between Rice Valley and the Colorado River. Because it has yet to be proven that a cone of depression in the water table encircles Danby Dry Lake, such a pathway is possible. Thus, they conclude that "Danby Lake (sic) may have much less discharge and less storage capacity than claimed in the site evaluation documents, and ground water may flow continuously from Ward Valley to the Colorado River."

Route 2: Route 2 is similar to route 1, except that it assumes that once the ground water enters Rice Valley, it passes through a divide between Rice and Vidal Valleys and then flows to the Colorado River. The Wilshire group argued that route 2 is possible because of the relatively great depth to crystalline rocks beneath the topographic divide between Rice and Vidal Valleys. Even if ground water flows above the crystalline rocks, it can be at a low enough level that a topographic divide would not impede ground-water transport.

Route 3: This route assumes pathways through low-density volcanic rock and fractured high-density rock within the mountains that flank the east side of Ward Valley (between the Stepladder and Turtle Mountains). The ground water then enters the alluvium in Chemehuevi Valley. Ground water is observed in shallow wells in the Turtle Mountains. The Wilshire group stated that if these observations represent perched water, which is accumulated water held above the water table by layers of lower permeability in the unsaturated zone, there can still be a regional water table at a deeper level with a continuous slope from Ward Valley to the Colorado. If, however, the wells penetrate the regional water table rather than perched water, the topographic high on the water table, which the water levels denote, precludes existence of a pathway between Ward and Chemehuevi Valleys.

Routes 4 and 5: Both of these postulated routes pass beneath the Sacramento Mountains, through fractured rocks above the detachment zone. The depth to the detachment zone here is uncertain. Route 4 requires a reversal of the present-day hydraulic gradient beneath the site, which would result in the ground water flowing in a direction opposite the known slope and direction of flow of the ground water (presumably by ground-water withdrawals in Piute Valley). Route 5 is presented as two possibilities, one through fractured upper plate rocks connecting Ward and Piute Valleys, the other is inferred because of the occurrence of a northwest-striking fault that crosses between the Stepladder and Sacramento

Mountains. The Wilshire group noted that springs high on the north side of the Sacramento Mountains would have to represent small perched ground-water zones rather than the regional ground-water system if this route is to be viable because their pathway assumes ground-water flow at a much deeper level than that of the springs.

The Wilshire group concluded from these arguments that several potential paths for ground-water flow through basin fill or fractured bedrock exist and must be tested for a complete analysis (Wilshire et al., 1994). They noted that although ground-water travel times are hundreds to thousands of years for these routes, the long-lived radionuclides slated for disposal at Ward Valley will be dangerous for much longer periods of time. It is their opinion that a full analysis of these potential pathways is required and should involve additional geophysical exploration, drilling, water chemistry testing, and ground-water flow rate determinations.

THE DHS/U.S. ECOLOGY POSITION

According to the Department of Health Services, the postulated hydrologic connections between the local ground-water flow system in Ward Valley and the Colorado River are not credible. Furthermore, these connections are not relevant to an objective evaluation of the Ward Valley low-level radioactive waste disposal site (Brandt, 1994a). DHS argues that, given a very low probability that dissolved radionuclides in the alluvial aquifer under the disposal site will ever reach the Colorado River, and because such a release would lead to no discernable radiological impact on the Colorado River, there is no risk to the public or to the environment.

The DHS described the shallow ground-water flow system beneath Ward Valley as "a relatively undynamic flow system". This characterization is based on the apparent magnitude of the north-to-south hydraulic gradient (3.0 meters per kilometer reported by US Ecology in the license application), and the estimated age of the ground water beneath the site (>10,000 years, based on carbon-14 dating).

Although not stated explicitly in their October 7 document, the DHS appears to accept the interpretation of recharge and regional ground-water flow that was proposed by M. Bedinger in his presentation to our panel, and as outlined in his USGS report (Bedinger et al., 1989). Precipitation falling in the mountains surrounding Ward Valley above an elevation of 1200 meters recharges a regional bedrock ground-water flow system. Bedinger postulates that this bedrock flow system may discharge to the Colorado River. His calculations of ground-water travel time in this regional system through bedrock are on the order of a million years (Bedinger et al., 1989). The alluvial aquifer in Ward Valley is a local ground-water flow system, possibly with underflow from Lanfair Valley, mountain-front recharge in the upper reaches of alluvial fans on the east and west sides of Ward Valley, and infrequent, localized recharge along Homer Wash. Local ground-water flow in Ward Valley is generally north-to-south, with discharge by evaporation at Danby Dry Lake.

Route 1 and 2: The DHS has stated, based on its review of the license application, that little, if any, ground-water flows in the alluvial deposits from Ward Valley to Rice Valley. Any such flow would be insignificant were it to occur. It is their position that, although direct evidence is insufficient to demonstrate conclusively the presence, position, and elevation of a ground-water divide between Ward and Rice Valleys, multiple lines of indirect evidence support the interpretation of a shallow ground-water divide. For example, they noted that the Wilshire group's postulated alluvial aquifer connection between Ward and Rice Valleys is contradicted by differences in the chemical composition of the ground water in southern Ward Valley and Rice Valley.

The DHS believes that additional information on the local ground-water flow system in Ward Valley or the ground-water divide between Ward and Rice Valley is not needed for purposes of evaluating the performance of the disposal facility. The saturated zone is viewed as providing an unnecessary but redundant isolation capability to that provided by the unsaturated zone under the disposal site. In the DHS view, potential transport of small concentrations of dissolved radionuclides in the local ground-water system from the northern portion of Ward Valley over 68 km southward to discharge at Danby Dry Lake would be so slow that it would allow substantial time for decay of the radionuclides to stable forms and provide ample opportunity for further reduction in the concentration of any dissolved radionuclides by dilution, dispersion, sorption, or precipitation. The DHS argues that another approximately 68 km of subsurface flow through Rice or Vidal Valley, if it were possible, would only increase confidence that this potential pathway would not adversely impact public health and safety or the environment.

Routes 3 and 5: DHS maintains that potential pathways through the mountains southeast and east of the site to Chemehuevi Valley (route 3) and Sacramento Wash (route 5) are contradicted by available data. The mountains on the east and west sides of Ward Valley represent recharge areas and ground-water divides for the regional and local ground-water flow systems. They pointed out that the Wilshire group "must disclaim existing water level data in wells and mine shafts in the mountains ..... in order to claim the potential for ground-water flow along these routes" (Brandt, 1994a).

Route 4: Route 4 is considered to be not credible by DHS and their contractors because: (1) it requires ground-water flow into one side of a hypothetical cone of depression, and ground-water flow out of a different side of that cone of depression, (2) it ignores the 135 years of planned operational and post-closure environmental monitoring and institutional control, and (3) it ignores evidence of low well yields in the Ward Valley area, and the large pumping lifts that would be required to develop the ground-water resource that could lead to large enough ground-water withdrawals to reverse the flow direction.

THE COMMITTEE'S APPROACH

The Geological Framework for the Ground-Water Pathways

An evaluation of the issue of ground-water pathways involves consideration of the bedrock structure beneath the alluvium, and the hydraulic conductivities of the older (Tertiary) alluvium, and of the upper and lower plate rocks. The configuration of the bedrock surface and whether or not bedrock highs beneath the alluvium rise above the present water table bear directly on the validity of some of the postulated pathways to the Colorado River.

The tectonic setting of the proposed Ward Valley LLRW site has been described by Wilshire et al. (1994), and is summarized briefly in Chapter 2 of this report. An important structure of concern is a Cenozoic normal or low-angle detachment fault between an upper plate of unmetamorphosed Mesozoic and Tertiary volcanic and sedimentary rocks, and a lower plate of metamorphosed older rocks. Partial answers to questions related to the nature and extent of interactions between the water table, the saturated zone, and bedrock materials and structures, and their bearing on the alternative ground-water pathways may be obtained using available geologic and geophysical data.

Subsurface Bedrock Topography

The series of geologic events outlined in Chapter 2 of this report reveals a scenario whereby a very irregular bedrock topography with considerable relief could exist throughout Ward Valley beneath the Quaternary alluvium (Figures 2.10 and 2.11). This irregular bedrock topography could influence the present-day ground-water regime.

Fracture Permeability

Bedrock types—both upper and lower plate rocks, with the exception of the old alluvium in the vicinity of Ward Valley—generally have a low porosity and permeability. The rates of ground-water flow in the bedrock will depend to a large extent on the degree to which fractures enhance the permeability of these rock units. Even within zones of enhanced fracturing, field experience at other sites suggests that the permeability will be lower than that of the Quaternary alluvium. This is generally the case in most of the Basin and Range, except in areas underlain by carbonate rocks (Bedinger et al, 1989). Where bedrock highs intersect the surface of the water table, as possibly occurs in some of the topographic divides along the alternative pathways proposed by the Wilshire group, additional opportunities exist for interaction between the ground-water system in alluvium and bedrock. The degree to which bedrock highs may perturb, or restrict, potential ground-water pathways in the alluvium will depend upon the height to which they penetrate the alluvium and the degree to which fracturing enhances the permeability of the bedrock units.

Prior to the August meeting, several members of the committee accompanied Keith Howard to a locality where a major detachment fault is exposed in the southern Piute Mountains. This fault separates high grade metamorphic rocks (Precambrian basement) of the lower plate from unmetamorphosed Miocene(?) sandstone and conglomerate ("old" alluvium). All of the rocks near the fault are intensely fractured and the fractures are open. Most often such fractures are closed at depth, either by mineral filling or by the overburden pressure (Pollard and Aydin, 1988). It is not known if all fractures are closed at depth in bedrock, but the spacing and intensity clearly change as a function of proximity to faults.

A Hydrologic Evaluation of the Proposed Ground-Water Pathways

The hydrogeologic database is inadequate to determine whether or not any component of the ground water flowing beneath the Ward Valley site eventually reaches the Colorado River. In order to evaluate potential pathways, it would be necessary to know: (a) the elevations of the water table along the potential pathways to determine if a gradient is sufficient for flow, (b) the three-dimensional distribution of fluid potential in the vicinity of any ground-water divide, and (c) the hydraulic conductivity of the geologic materials along the proposed pathway to determine if those materials have sufficient permeability to allow significant amounts of water to be transmitted through them.

Water-Level Data

Water-level data at the regional scale are limited, both spatially and temporally (LA Appendix 2420.A). The paucity of wells, particularly of wells with more than one or two recorded water level measurements, does not permit the accurate determination of regional hydraulic gradients and flow directions. The best-supported hydraulic gradient, that through the Ward Valley alluvium, is based on water-level measurements in four monitoring wells. Many of the data for the older wells are from the first half of the century, are poorly documented, and are of unknown reliability.

Hydraulic Conductivity

Hydraulic conductivity determinations exist only for the alluvium beneath the proposed LLRW site, and those are of questionable quality. Because no observation wells were within the area of influence of the pumping wells during aquifer testing, all the pumping test analyses were interpreted as single well tests. Single well tests generally underestimate transmissivity because well losses produce excess drawdown over that which occurs in the porous medium surrounding the borehole. The range of values for hydraulic conductivity obtained with the single-well pumping tests is 0.02 to 0.7 m per day, with a geometric mean of 0.12 m per day (LA Section 2420.1.3.1). David Huntley, consultant to the Wilshire Group, estimated hydraulic conductivities to be on the order of 1.2 to 12.2 m per day (Huntley, 1994). In their hydrogeologic review of the Ward Valley site prepared for the Metropolitan Water District of Southern California, GEOSCIENCES Support Services, Inc. used a range of 2 to 24 m per day in their calculations; this range was based on various types of data, including grain-size analyses, analyses of geophysical logs, and pumping test analyses from adjacent valleys (GEOSCIENCES Support Services, Inc., 1994).

It is also possible to estimate the hydraulic conductivity of the alluvial sediments in Ward Valley on the basis of a calculation using Darcy's law (Equation 3.1 in Chapter 3 of this report), and assuming that most of the ground water flowing through the valley discharges at Danby Dry Lake (see Box 5.1). On this basis, the committee estimated the hydraulic conductivity of the alluvial valley fill to range from 2.7 to 6.1 m per day. These values are an order of magnitude greater than the values used in the license application, are similar to those used by the Wilshire group, and are slightly less than the values estimated by GEOSCIENCES Support Services, Inc. The ground-water travel times to Danby Dry Lake, determined with our estimates of pore-water velocities, range from 850 to 1800 years (Box 5.1). These estimates can be compared with the age of the ground water beneath the Ward Valley site, estimated using ^{14}C dates (See Chapter 3). They are consistent, in a general sense, if it is assumed that the ground water beneath the Ward Valley site originated from recharge areas at the northern end of Ward Valley or within Lanfair Valley.

Beyond the boundaries of the site, there are neither hydraulic conductivity measurements for the basin-fill alluvium within Ward Valley, nor information on the nature and variability of sediment type along the length of the valley that might serve to estimate hydraulic conductivity. However, the uplands forming the valley boundaries, from which the valley-fill sediments are derived, are similar along the entire length of the valley. Therefore, the physical characteristics of the valley-fill alluvial sediments are not expected to vary greatly over most of the valley length. The nature of the fracture permeability in the bedrock beneath and surrounding the site is unknown, particularly the extent of fracturing and the degree to which those fractures may have been healed by the deposition of minerals along their openings (See discussion in Chapter 3 on fracture permeability).

The majority of the ground-water flow beneath the site appears to go to Danby Dry Lake, based on limited hydrologic and geologic data and topographic conditions. The extensive efflorescent deposits indicating continuous, long-term evaporation from a shallow capillary fringe at Danby Dry Lake, the shallow level of the ground water (approximately 1.5 m below the playa surface), and the high salinity of the ground water attest to the fact that evaporation is occurring from the playa surface and that this is a major discharge area. If a large portion of the discharge were occurring elsewhere, it would be unlikely that the shallow water table would coincide with the playa.

It cannot be ruled out, however, with the presently available data, that some portion of the ground-water may be leaving the basin. While Ward Valley may be an enclosed surface-water basin, it is not necessarily enclosed with respect to ground water. Some combination of permeability and hydraulic gradient for a particular pathway could permit a fraction of the flow from the valley to exit towards the Colorado River. Such pathways could be either

Box 5.1

Estimated Hydraulic Conductivity and Travel Times for Ward Valley

An estimate of the hydraulic conductivity of the alluvial sediments in Ward Valley can be made using Darcy's Law, a means of calculating the rate of movement of ground water, and assuming that most of the flow through the valley discharges at Danby Dry Lake. Estimates of discharge at the lake vary from 1.4×10^7 m^3/yr (11,000 acre-ft/yr) (Law/Crandall, 1992), based on a fraction of the pan evaporation and the area of salt-covered sediments, to 2.7×10^7 m^3/yr (22,000 acre-ft/yr) (LA Section 2420.1.2.2), based on potential evaporation rates and the surface area from which evaporation could occur. In the Law/Crandall calculation, an evaporation rate of 24 cm/yr was used, which is greater than published values in the recent literature. Allison and Barnes (1985) found an average evaporation rate from a dry lake to be 17 cm/yr, while Ullman (1985) determined 1 to 2 cm/yr for long-term evaporation rates from a salt-crust covered playa.

If one considers the Law/Crandall estimate as a reasonable upper bound and takes into account that areas south and west of Danby Dry Lake, that are not part of Ward Valley, also contribute ground water to the discharge at Danby Dry Lake, a contribution of 1.1×10^7 m^3/yr (9,000 acre-ft/year) is a reasonable upper bound on the discharge (Q) through Ward Valley. If the lower evaporation rates cited from the literature are more accurate, a value of 4.9×10^6 m^3/yr (4,000 acre-ft/yr) would be more reasonable. The hydraulic gradient (dh/dl) as measured from the Camino Well (ground-water elevation, 437 m above mean sea level (amsl)) to Danby Dry Lake (ground-water elevation, 184 m amsl) over a 68 km distance is 0.0037. An average saturated cross-sectional area for the valley was estimated to be 250 m thick and 5.6 km wide. The effective porosity (n_e) for the partially cemented alluvial materials is assumed to be 0.10 (LA Section 2420.1.3.2; GEOSCIENCE Support Services, Inc., 1994; Huntley *in* Wilshire et al., 1994).

Using Darcy's law (see Equation 3.1, after replacing q by Q/A and H by h for horizontal flow), the hydraulic conductivity (K) can be calculated:

$$K = Q/ (A\ dh/dl)$$
$$K = 4.9 \times 10^6 \text{ m}^3/\text{yr} / (250 \text{ m} \times 5600 \text{ m} \times 0.0037)$$
$$= 946 \text{ m/yr} = 2.6 \text{ m/d}$$

For a Ward Valley contribution to evaporation at Danby Dry Lake of 1.1×10^7 m^3/yr, the hydraulic conductivity would be calculated to be 5.8 m/d.

Corresponding pore water velocities (v_p) are:

$$v_p = (K/n_e)\ (dh/dl) = 2.6 \text{ m/d} / 0.10)\ (0.0037) = 9.6 \times 10^{-2} \text{ m/d} = 35 \text{ m/yr}$$

for the lower discharge rate and 0.22 m/d (= 79 m/yr) for the higher discharge rate.

Travel times from the proposed site to Danby Dry Lake over the 68 km distance (L) are:

$$L / v_p = 68,000 \text{ m} / 35 \text{ m/yr} = 1900 \text{ yr}$$ for the lower discharge rate and 860 yr for the higher discharge rate.

through (a) continuous alluvial deposits through adjacent valleys, or (b) a combination of alluvial and bedrock pathways through the adjacent mountains and valleys, as described previously in this section.

The hydrogeology of Miocene extensional valleys of southeastern California and the Great Basin of Nevada and Utah is controlled in part by the permeability contrast between the alluvial valley-fill deposits and the bedrock. The other major hydrologic controls are the spatial and temporal distribution of recharge. Higher precipitation and often higher recharge rates associated with higher elevations of the mountains generally (but not always) create ground-water divides between adjacent valleys. These factors will determine whether the flow is restricted to a local basin flow system or if interbasin regional flow occurs. The license application presumes that the surrounding and underlying rocks are of relatively low permeability and that ground-water flow divides are associated with the topographic divides, thus creating an enclosed ground-water basin as well as an enclosed surface-water basin. Flow in the carbonate rock province of the Great Basin, which includes much of Utah and Nevada and extends into California to a point about 80 km north of the Ward Valley site, is characterized by interbasin movement of ground water through the permeable lower carbonate aquifer (Winograd and Thordarson, 1975). In locations such as Yucca and Frenchman Flats the ground water in the younger hydrogeologic units moves principally downward into the underlying lower carbonate aquifer. Elsewhere, such as in the southern Amargosa Desert and in southern Indian Springs Valley, the ground water in the younger sediment derives from upward leakage of water from the underlying lower carbonate aquifer. Permeability in this lower carbonate aquifer is not related to primary porosity, but to fracture permeability enhanced by dissolution. There is no carbonate bedrock between the Ward Valley site and the Colorado River.

Winograd and Thordarson (1975) stated that, with reference to downward hydraulic gradients, "declining heads with increasing depths are attributed to at least three causes: (1) penetration of an unsaturated zone beneath a perched zone; (2) energy losses in areas of dominantly vertical movement, as is common in recharge areas or areas with semi-perched ground water; and (3) penetration of an aquifer of the same or greater transmissibility than the overlying aquifer but with a lower discharge point."

While the Ward Valley site is clearly outside of the carbonate aquifer province, a downward vertical hydraulic gradient is apparent between MW-01 and MW-02. It is greater than the horizontal hydraulic gradient by approximately a factor of three. The explanation for this apparent downward hydraulic gradient is unclear. One hypothesis (Harding Lawson Associates, 1994) is that more transmissive alluvial deposits exist deeper in the aquifer system and that such a depositional pattern would be expected in the valley. The permeameter determinations revealed lower hydraulic conductivities for a number of the deeper core samples (LA Sections 2420.2.1.2 and 2420.3.1), indicating that if the valley does contain alluvium with higher permeability, it must be below the current depth of investigation.

Another possible explanation for the apparent downward hydraulic gradient is the presence of a higher permeability zone in the fractured fault blocks beneath the site and along the surface of the detachment fault itself. If highly fractured fault blocks exist, they could have significant permeability, as could the detachment fault zone.

Still another possible explanation of the apparent vertical gradient is non-verticality of either or both wells involved, as mentioned in Chapter 3.

REVIEW OF PATHWAYS

All pathways postulated by the Wilshire group are predicated on the fact that the Colorado River (and immediately adjacent ground water) is approximately 300 m lower than the water table beneath the site. If no intervening water table highs or low permeability hydraulic barriers exist, which would create a ground-water flow divide, ground water could flow towards the lower elevation of the Colorado River.

Routes 1 and 2 (Rice Valley and Vidal Valley): These routes are possible because of an apparent continuity of permeable alluvial materials from Ward Valley through Rice and Vidal Valleys. Arguments that a ground-water divide exists beneath the topographic divide in the region leading to Rice Valley are not supported by any reliable water level data. Well-survey data (LA Figure 2420.A-4) indicate water levels in the area that are lower than the water levels at Danby Dry Lake, therefore indicating a hydraulic gradient that could drive flow towards Rice and Vidal Valleys. Because the water-level data date back three to eight decades, however, their reliability is uncertain. Although no information precludes a ground-water divide, its presence cannot be substantiated with the current data.

Arguments against a flowpath connecting Ward and Rice Valleys based on differing major ion chemistries draw upon observations in Thompson et al. (1987) which suggested sodium sulfate water in Rice Valley as opposed to sodium bicarbonate water in Ward Valley. The determination for Rice Valley is based on a single sampling point, and no sampling point is indicated for Ward Valley in that work. This database is insufficient to allow us to draw reliable conclusions. Bedinger et al. (1989) indicated that both Ward Valley and Rice Valley contain primarily sodium bicarbonate water. Without further corroboration, the geochemical argument is not defensible.

The fraction of flow from Ward Valley, if any, which discharges into Rice and Vidal Valleys as opposed to evaporating at Danby Dry Lake, is unknown and, therefore, cannot be categorically considered insignificant, as was done by DHS. However, long travel times, on the order of thousands of years, and long travel distances to the Colorado should minimize adverse effects of potential releases from the proposed site.

Route 3 (Stepladder Mountains to Chemehuevi Valley) and Route 5 (Sacramento Mountains): These possible pathways would pass through volcanic or crystalline upper-plate rocks. If these rocks contain open fractures, they could be locally permeable, although their permeability is thought to be less than that of the alluvium in Ward Valley. No hydraulic conductivity data exist for these rocks, although partial healing of the fractures by the deposition of hydrothermal minerals could be expected to reduce the permeability.

Bedinger et al. (1989) estimated the relative ground-water velocities at the water table in the rocks to be 2 to 3 percent of the velocities in the alluvium (Plate 4 in Bedinger et al., 1989). They also modeled travel times through the volcanic rocks of the Turtle Mountains and then through Chemehuevi Valley to the Colorado River to be on the order of tens of thousands to hundreds of thousands of years (Plate 6 in Bedinger et al., 1989).

High water levels have been noted in springs and wells within the mountains. These may reflect local perched ground-water systems or they may reflect the shallow regional water table. The presence of a shallow hydraulic gradient from the mountains westward towards Ward Valley, if such a gradient exists, does not eliminate the possibility of eastward flow in a deeper flow system. This is particularly true if the deeper flow system is partially isolated from the shallow system, such as could occur with a heavily fractured zone along the detachment fault.

Route 4 (Sacramento Wash to Piute Valley): The plausibility of this pathway is predicated on long-term, large-volume pumping of ground water in the Piute Valley. The pumping could presumably lower the water table over a sufficiently large area to reverse the hydraulic gradient at the site and cause the direction of ground-water flow to change from southward to northward. The likelihood of such pumping is unknown. The pathway would pass through upper plate rocks which would be permeable if fractured, although presumably less permeable than the alluvium. While not impossible, this potential pathway is speculative and unlikely, in the committee's judgement.

DHS argues that monitoring will continue for 135 years, and this should detect any such reversal of the hydraulic gradient. They do not address what would happen if such a reversal were to occur after that time period. They also argue that the path requires flow into one side of the pumping cone of depression and out the other side, which is not plausible. Furthermore, in the committee's opinion, if high activities of radionuclides were detected in the hypothetical wells in Piute Valley, the wells would most likely be turned off and some kind of remedial action implemented.

Another potential impediment to ground water flow from beneath the site eastward to the Colorado River along the postulated pathways discussed in this section is influence of potential recharge from Homer Wash. If sufficient recharge occurs from this wash, the hydraulic effects may be sufficient to confine ground water in the valley fill west of the wash to the west side of the valley, thus preventing significant flow paths from crossing Homer Wash.

We would have a better understanding of the alternative regional pathways, gradients, ground-water flow rates, etc., if data from three drill holes to bedrock were available: (1) midway down Ward Valley; (2) on the Rice-Ward Valley topographic divide; and (3) on Route 3 through the Stepladder Mountains. **In the judgement of the committee, however, it would not be possible under any reasonable scenario for site characterization to confirm, or eliminate with absolute certainty, the postulated regional bedrock pathways.**

THE MAGNITUDE OF POTENTIAL IMPACTS OF LONG-LIVED RADIONUCLIDE MIGRATION FROM THE SITE TO THE COLORADO RIVER

Although it is the committee's view that most, if not all, of the Ward Valley ground water discharges at Danby Dry Lake and evaporates, a portion of the water could possibly bypass Danby Dry Lake through one or more of the four pathways the committee regards as possible to reach the Colorado River, Rice Valley (Route 1), Vidal Valley (Route 2), the Stepladder Mountains (Route 3), or the Sacramento Mountains (Route 5). To assess the relative significance of these scenarios, one can make some bounding estimates of potential impacts on the water quality of the Colorado River using one of the pathways.

Assumptions for Bounding Estimates

To make such bounding assessments, some initial assumptions are needed regarding basic flow system characteristics, particularly the approximate path lengths and approximate travel times.

Path Length

Based on Figure 5.1 (from Wilshire et al., 1993) showing postulated pathways to the Colorado River, routes 1 and 2 are approximately 130 km long (through the Rice or Vidal Valleys). The other possible pathway is the Wilshire group's route 3 eastward beneath the hills at the south end of the Stepladder Mountains. The approximate length of that pathway is 80 km from the proposed site to the Colorado River.

Travel Time

As discussed in the previous section, sufficient information is available to allow reasonable estimates of ground-water travel times to be made with some confidence for the purpose of this analysis.

As indicated above, the gradient along the Ward Valley portion of routes 1 and 2 is estimated to be about 0.0037 m/m, based on water levels in the monitoring wells at the proposed site and the elevation of Danby Dry Lake. The overall change in hydraulic head to the Colorado River is about 360 m in 130 km for a gradient of 0.0028 m/m. Travel time can be estimated using the Darcy velocity equation (page 5-16), together with estimates of hydraulic conductivity, gradient and effective porosity. Assuming the total pathway gradient is 0.003 m/m and K = 5.8 m/d, v_p = (5.8 x 0.003)/0.1 = 0.18 m/d or 67 m/yr. For K = 2.6 m/d, the velocity is 0.078 m/d or 28 m/yr. Total travel time along routes 1 or 2 of 130 km would then be about 1,900 years, for K = 5.8 m/d and 4,600 years for K = 2.6 m/yr. These are adequate times for the decay of short-lived fission and activation-product radionuclides present in low-level waste; however, they are not sufficient times for the decay of long-lived alpha-emitting transuranics such as plutonium-239 to decay (half life = 24,100 years). This travel time analysis assumes that the plutonium reaches the water table immediately upon emplacement in the facility and allows no time for water and dissolved radionuclides to migrate vertically downward from the site through the unsaturated zone to the aquifer. As discussed in Chapter 3, it is expected that the residence time of any radionuclide migrating through the unsaturated zone will be at least on the order of hundreds of years.

Thus, for the purpose of this bounding assessment, it becomes important only to examine the potential impacts of waste components having half-lives longer than a few hundred years.

Potential Impacts on the Colorado River

Potential Radioactive Contamination by Ground Water

Plutonium is one of the longer-lived components of low-level waste that can be examined. DHS estimates that the total plutonium inventory of all the waste received by the site over its 30-year operational history will be from "a fraction of a curie to 2 curies from decontamination waste, and a fraction of one curie to several curies from other sources" (Brandt, 1994b). Radionuclide composition of the waste is certified by the generators and will be monitored by the DHS and the site operator. DHS has enforcement authority over radionuclide quantity restrictions.

As discussed in Chapter 1, based on the report from the Congressional Research Service (Holt, 1994) and the letters from the US NRC (1994a,b) addressing the controversy over the amount of ^{239}Pu expected among the wastes, the committee applied the lower order of magnitude agreed upon by both agencies for the estimate. If we assume that "several" plus two curies means approximately 10 curies, then a potential bounding impact can be computed.

For a conservative estimate, we can assume a scenario in which the entire 10 curies of plutonium is released into the underlying aquifer over a period of 30 years (the same rate at which it enters the repository). This release scenario is not credible in the Committee's opinion, but it can serve to introduce a degree of conservatism for this calculation. We further assume that the plutonium is released in a chemical form that is some sort of stable organic complex, (such as a chelated form, like ethyldiamine tetra-acetic acid (EDTA), a commonly used chelating agent that is present in part of the waste). Chelated plutonium is not subject to solubility constraints nor any retardation reactions, such as adsorption, which adds further conservative assumptions. In other words, it migrates along the ground-water flow pathway at the same speed as the ground-water pore velocity and is therefore not subject to any significant decrease of concentration through retardation and radioactive decay. This assumption, like the first, is also not credible in our judgement, but nonetheless can be useful in developing a conservative calculation.

Under this scenario, one thirtieth of the assumed 10 curies (Ci) plutonium inventory, or 0.33 curies per year (Ci/yr), would enter the Colorado River each year for 30 years, once it traversed the path length. According to the Metropolitan Water District (Fisher, 1992), the average flow of the Colorado River below Needles is about 10 billion m^3/yr (8 million acre-feet per year). In the future this flow is likely to decline significantly as the up-river states divert more of their full legally appropriated shares. Therefore, in this analysis we assume the average flow to be 5 billion m^3/yr (4 million acre-feet per year), which is about the flow rate required to satisfy down-stream legal allocations. If 0.33 Ci of plutonium entered a river flow of 4 million acre-feet per year, the resulting concentration would be 0.07 pCi/l (4 million acre ft = 4.9×10^{12} liters and 1 Ci = 10^{12} pCi). According to the Metropolitan Water District (MWD), (Fisher, 1992) the average present concentration of natural alpha-emitting isotopes in Colorado River water is about 4.4 pCi/l, based on recent river flow rates which are much higher than the future flow rate assumed for the above plutonium calculation. **Therefore, an addition of 0.07 pCi/l would be insignificant, in the committee's judgement, especially when compared to the MWD's health-based regulatory standard for gross alpha-emitting radionuclides of 15 pCi/l (which excludes the natural isotopes of uranium, ^{226}Ra , and ^{222}Ra). In other words, the river is currently transporting 44 Ci per year of natural alpha-emitting isotopes. The addition of 0.33 Ci per year is small compared to the annual natural load of 44 Ci per year.**

A few other long-lived alpha emitting radionuclides may also be present in some of the waste (i.e., Uranium-238 and Americium-234) but in estimated quantities much lower than plutonium. The foregoing analysis for plutonium would therefore be applicable and relevant also for those other long-lived isotopes.

It is also important to note that this hypothetical discharge of 0.33 Ci/yr from the site to the Colorado River would require a combination of circumstances that has an incredibly low probability of occurring. This calculation is presented, therefore, only to show that actual potential impacts, if any, would be much less than the result calculated here. **The committee based the calculation on an assumed waste plutonium inventory of 10 Ci. Even if the released plutonium quantity were 100 times greater than assumed here (10^3 or 1000 curies), the maximum hypothetical impact on the Colorado River concentrations would be 7 pCi/l, which approaches but still remains less than the health-based regulatory criterion of 15 pCi/l. Implicit in this worst-case scenario is the assumption of instantaneous transport of radionuclides from the disposal site to the water table, giving no credit for retardation or adsorption in the unsaturated zone. If, however, future analysis indicates that significantly greater amounts of plutonium will be disposed in the facility, more detailed modeling incorporating the transport processes affecting the concentration of plutonium would have to be performed.**

Factors Affecting Radionuclide Transport

Some of the key factors that would greatly diminish potential impacts of long-lived isotopes on the Colorado River are the following:

1. Because much of the Ward Valley ground water likely discharges at Danby Dry Lake, it is not credible for the entire quantity of any isotopes released from the site into the aquifer to follow a minority pathway (if it exists) to the river.

2. It is unlikely that plutonium or any other transuranic isotopes could remain in a highly soluble and unretarded chemical form for the entire duration of the migration to the river. The most mobile and frequently employed complexes of plutonium are organic chelates, which tend to be unstable and degradable over long periods of time (Means and Alexander, 1980). Therefore, significant retardation would ultimately cause great delays in plutonium migration and concurrent decay.

3. No credible mechanism has been identified that could dissolve and leach out all the plutonium (or any other isotope) from the site over a period of a few decades and carry it to the aquifer. Therefore, releases, if any, to the aquifer would be much lower than assumed.

4. Even if the plutonium were released over a few decades, dispersion processes over the long pathways would spread the plume out over a long period of time and dilute it to an extent that any potential impacts on the Colorado River would be significantly diminished below that calculated herein.

Conclusion Concerning Contamination of the Colorado River

Although the committee considers that there are conceivable, but unlikely, flow paths for some of the Ward Valley ground water to the Colorado River, we conclude that the potential impacts on the river water quality would be insignificant relative to present natural levels in the river and to accepted regulatory health standards.

Potential Radioactive Contamination by Airborne Dust

Another potential scenario that has been postulated for impacting the Colorado River or the water in the Colorado River Aqueduct is by airborne transport of radioactive dust from Danby Dry Lake. Under this scenario, plutonium would be leached out of the site into the aquifer, and carried to Danby Dry Lake where it would be deposited as an evaporite precipitate into the sediment as the ground water evaporated. It would then be picked up by wind and carried eastward to the Colorado River.

If we again allow all of the assumed 10 Ci plutonium inventory to be released over 30 years (0.33 Ci/yr), it will be deposited into the Danby Dry Lake bed at the same rate some time later (again assuming no retardation or dispersion). Although the Law/Crandall (1992) study of Ward Valley concluded that evaporation at Danby Dry Lake occurs over an area of 75 km^2, we assume, for conservative purposes, that the major evaporation area is only 21 km^2 based on information presented by J. P. Calzia (1992). If the plutonium is deposited over this area in the upper inch of soil, it would amount to 0.33 Ci per 19×10^6 ft^3 of soil per year, which is about 0.33 Ci per 4×10^9 kg, or about 0.08 pCi/g. Only a small fraction of that soil could be transported away from Danby Dry Lake by wind each year, but even if it all were transported, only a small fraction could fall out specifically into the Colorado River or the Colorado River Aqueduct, which occupy relatively small areas. In this case a combination of circumstances with an extremely low probability would have to occur (just as in the previous scenarios) to cause any substantial portion of the long-lived isotopes to become deposited in the Danby Dry Lake sediments. **Therefore, the committee concludes that the potential impact of airborne radionuclides on the water quality of the Colorado River or Colorado River Aqueduct is minuscule, even much less than the upper bounding calculation made previously for the direct ground-water pathway scenario to the river.**

SUMMARY OF CONCLUSIONS

The committee concludes that:

1. Based on limited hydrologic and geologic data and the topographic conditions, the majority of the ground-water flow beneath the proposed site in Ward Valley appears to discharge at Danby Dry Lake. It cannot be ruled out, however, that some portion of the ground water passing beneath the proposed site may leave the Ward Valley basin.

2. It would not be possible under any reasonable expectation for site characterization either to confirm, or to eliminate with absolute certainty, any of the regional bedrock pathways that have been postulated by the Wilshire group.

3. While there are conceivable, but unlikely, flowpaths for some ground water within Ward Valley to reach the Colorado River, conservative bounding calculations suggest that the potential impacts on the river water quality would be insignificant relative to present natural levels of radionuclides in the river and would meet accepted regulatory health standards.

4. The committee concludes that the potential impact of airborne radionuclides on the water quality in the Colorado River or Colorado River Aqueduct is minuscule, much less than the upper bounding calculation for the direct ground-water pathway scenario to the river.

REFERENCES

Bedinger, M. S., Sargent, K. A., Langer, W. H. 1989. Studies of the geology and hydrology of the Basin and Range province, southwestern United States, for isolation of high-level radioactive waste - characterization of the Sonoran region, California. U.S. Geological Survey Professional Paper 1370-E, 30 p.

Brandt, E. C. 1994a. Summary of the California Department of Health Services presentation to the Committee to Review Specific Scientific and Technical Issues Related to the Siting of a Proposed Low Level Radioactive Waste Disposal Facility in Ward Valley, California. October 6, 1994.

Brandt, E.C. 1994b. Letter to I. Alterman, September 22.

Calzia, J.P. 1992. Geology and saline resources of Danby Lake Playa, southeastern California. Pp. 87-91 in Old Routes to the Colorado, Reynolds, R. E., compiler. San Bernardino County Museum Association Special Publication 92-2.

Fisher, S. M. 1992. The Metropolitan Water District of Southern California Annual Report for the Fiscal Year July 1, 1991 to June 30, 1992.

GEOSCIENCES Support Services, Inc. August 1994. Hydrogeologic review of the proposed Ward Valley Low-Level Radioactive Waste Facility, prepared for the Metropolitan Water District of Southern California.

Harding Lawson Associates. 1994. Summary of well installation, water-level monitoring, and aquifer testing, Ward Valley, California: Unpublished Report. October 6.

Holt, M. 1994. Plutonium Disposal Estimates for the Southwestern Low-Level Radioactive Waste Disposal Compact. Memorandum dated June 3. Congressional Research Service. Library of Congress. Washington,

Huntley, D. 1994. Directions and rates of groundwater flow, Ward Valley, in Ward Valley proposed low-level radioactive waste site: A report to the National Academy of Sciences. Wilshire, H. G., K. A. Howard, and D. M. Miller, eds.

Law/Crandall, Inc. 1992. Water resources evaluation, Ward Valley, San Bernardino County, California, prepared for City of Needles.

License Application. 1989. U.S. Ecology, Inc. Administrative Record, Ward Valley Low-Level Radioactive Waste Disposal Facility, Section 2420, and Appendix 2420.

Means, J. L. and C. A. Alexander. 1980. Final report on the environmental geochemistry of chelating agents and radionuclides complexes, BMI-X-701, Battelle Columbus Laboratories.

Pollard, D. D. and A. Aydin. 1988. Progress in understanding jointing over the past century. Geological Society of America Bulletin. 100:1181-1204.

Thompson, T. H., J. A. Nuter, W. R. Moyle, and L. R. Wolfenden. 1987. Maps showing distribution of dissolved solids and dominant chemical type in ground water, Basin and Range province, southern California. U.S. Geological Survey Water Resources Investigation Report WRI-83-4116C.

U.S. Nuclear Regulatory Commission. 1994a. Letter from M. R. Knapp to Ina Alterman, National Research Council. Dated November 23.

_____1994b. Letter from M.R. Knapp to Ina Alterman, National Research Council. Dated December 14.

Wilshire, H. G., K. A. Howard, and D. M. Miller. 1993. Description of earth-science concerns regarding the Ward Valley low-level radioactive waste site plan and evaluation.

Wilshire, H. G., K. A. Howard, and D. M. Miller. 1994. Ward Valley proposed low-level radioactive waste site: a report to the National Academy of Sciences.

Winograd, I. J., and W. Thordarson. 1975. Hydrogeologic and hydrochemical framework, south-central Great Basin, Nevada-California, with special reference to the Nevada Test Site, U.S. Geological Survey Professional Paper 712-C.

6

SUBSURFACE MONITORING PROGRAM

Issue 4

Adequacy of Plans to Monitor
Ward Valley Ground Water and Unsaturated Zone

THE WILSHIRE GROUP POSITION

The specific issues raised by the Wilshire group were the lack of plans for post-closure monitoring of the unsaturated zone or the ground water downgradient from the site and the absence of remediation plans if unacceptable levels of radionuclides are detected off site. In their first memorandum, based on the draft EIR/S, the Wilshire group asserted that no plans were in place for post-closure monitoring (Wilshire et al., 1993a). Subsequently, upon reading more recent licensing documents, they questioned the adequacy of the plans that were in place (Wilshire et al., 1993b).

THE DHS/U.S. ECOLOGY POSITION

The DHS and U.S. Ecology pointed to the detailed plans to conduct monitoring of the unsaturated zone and the ground water during the operation of the facility ($\leq$ 30 yrs). These plans are described in a later section in this chapter. DHS/USE stated that once the disposal operations are completed, U.S. Ecology will conduct post-closure environmental monitoring for 5 years. This period of post-closure monitoring can be extended until DHS determines that the closure requirements have been met. The institutional control period ($\leq$ 100 yr) must include an environmental monitoring program according to 10 CFR Part 61 to provide early warning of radionuclide release.

Two basic types of monitoring are proposed for Ward Valley: (1) regulatory compliance monitoring to assure that contaminant releases do not exceed regulatory levels at the disposal system boundaries and (2) performance monitoring to provide an early warning of releases that may exceed regulatory levels. The compliance boundaries are the air, vegetation, and water table at the edge of the buffer zone. Monitoring in the unsaturated zone constitutes performance monitoring because the unsaturated zone is not a regulatory compliance boundary. Performance monitoring of the unsaturated zone is critical because this is the primary barrier to radionuclide migration. The data provided by the proposed unsaturated-zone monitoring program will be critical in evaluating the performance of the Ward Valley facility and will be compared with the results of performance assessment models of the site.

THE COMMITTEE'S APPROACH

Recognizing the importance of the unsaturated-zone monitoring, the committee emphasizes in this section performance monitoring that will be conducted to determine if the site is performing as expected.

THE PERFORMANCE MONITORING PLAN FOR THE UNSATURATED ZONE

Emphasis has generally been placed on performance assessment modeling to demonstrate the performance of near-surface waste disposal facilities. Such modeling is used to evaluate the long-term performance of engineered barriers and the natural system. Lack of field data to verify and validate the computer models, however, would limit certainty in model predictions and confidence in simulation results. As very few sites have been studied that can be considered analogs to the proposed engineered barrier system at Ward Valley, our understanding of how these systems perform is limited.

The committee also notes that the Ward Valley monitoring program does not specifically address the monitoring program relative to performance assessment during operation, closure, and post-closure site evaluations.

Scope and Purpose of a Monitoring Plan

Monitoring is defined in 10 CFR Part 61 as "observing and taking measurements to provide data to evaluate the performance and characteristics of the disposal site" (U.S. Code of Federal Regulations, 1982). This general definition of monitoring suggests that site characterization should be an integral part of a monitoring program and is an ongoing activity that continues during operation and the extended period of institutional control of the site. Although traditionally site characterization, monitoring, and performance assessment have each been conducted independently, various agencies involved in waste disposal such as the Nuclear Regulatory Commission (NRC) and the Department of Energy (DOE) increasingly recognize that site characterization, monitoring, and performance assessment should be integrated (Campbell et al, 1993).

Although fundamental site characterization data are collected prior to license application, it is the committee's judgment that site characterization should be continued through the operational phase (U.S. Code of Federal Regulations, 1982). Continued site characterization data collection and monitoring through operations will provide an understanding of long-term processes. Monitoring ambient conditions distant from the engineered facility will provide background data and act as a control for comparison with monitoring data from the engineered facility. An additional reason for continuing site characterization studies is that, as our conceptual understanding evolves of how arid systems behave and our ability to measure and monitor hydraulic parameters improves, the resulting reliable data will reduce uncertainties and increase confidence that the site will perform effectively in isolating the waste. Site characteristics are critical to waste disposal in arid settings because reliance is ultimately placed on the natural system to contain the waste. The site characterization and monitoring data collected during operation should be used for the performance assessment analysis required prior to closure. A continuing performance assessment can be used to guide site characterization and monitoring by determining what the critical data are and also by defining the optimal sampling locations and frequency. **In the opinion of the committee, integration of site characterization, monitoring, and performance assessment can be regarded as an iterative process with feedback from each component.**

Scope of Ward Valley Proposed Monitoring Plan

The applicant's proposed monitoring program outlined by U.S. Ecology (1991) does not include the integrated approach discussed above and is limited to monitoring only. The applicant plans to update the performance assessment if the monitoring results do not agree with the performance assessment (U.S. Ecology, 1991, Figure IV.b.a). The applicant's monitoring program includes up to 30 yr of operational monitoring, 1 yr of closure monitoring, 3 to 5 yr of post-closure monitoring and 100 yr of institutional monitoring.

PERFORMANCE MONITORING OF THE UNSATURATED ZONE

The original unsaturated-zone monitoring plan in the license application was subsequently enhanced on the basis of recommendations from a select committee on this subject. Thirteen objectives for the proposed monitoring program were outlined and included the following:

- provide early detection of significant contaminants;
- monitor beneath, adjacent to, and above the waste;
- incorporate improved monitoring technology as it becomes available;
- avoid compromising the integrity of the disposal system;
- provide acceptable accuracy and reproducibility.

The monitoring systems for the A and B/C trenches are similar. A critical component of the monitoring program is the concept of monitoring islands (2.4 to 3 m diameter) (Figure 6.1). The islands will be placed at approximately 60 m intervals that correspond to incremental excavation of the trenches. Three islands will be placed in each cross section: one in the center of the trench and two in the margins of the trench. Instrumentation similar to that installed in the islands will be installed in the undisturbed sediments adjacent to the trench. The islands will consist of vertically stacked concrete culverts that will be constructed in advance of the surrounding waste and backfill (Figure 6.1). To minimize damage during

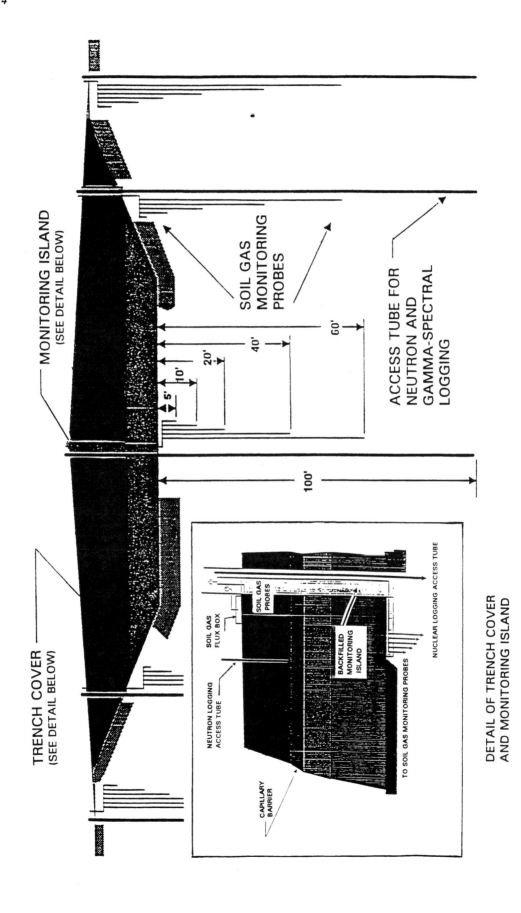

Figure 6.1 Schematic diagram of trench cover and monitoring islands. Inset: Detail of monitoring island (U.S. Ecology, 1991). (Distances not converted to metric system.)

operations, 2.7 m³ bulk soil bags will be used. The caissons may be left open while the site is active to allow access to instrumentation and will be backfilled during facility closure. A trench-cap demonstration unit (TCDU) will be constructed outside the disposal facility as part of the monitoring program.

Monitoring the Trench Cover

The main objectives of monitoring the trench cover are:

- identify and quantify releases of volatile radionuclides such as ^{3}H, ^{14}C, and radon (^{222}Rn) to the atmosphere and vegetation;
  - quantify concentrations and concentration gradients of volatile radionuclides;
  - evaluate the performance of the cover in minimizing infiltration.

To meet these objectives, the monitoring will consist of:

- measurement of vertical water-vapor flux from the land surface using flux chamber, porometer, or other equivalent method;
  - measurement of plant transpiration using flux chamber, porometer, or other equivalent method;
  - vegetation sampling and analysis for ^{3}H and ^{14}C radionuclides;
  - soil-gas sampling at 1.5, 3.6, and 5 m semiannually and analysis of ^{3}H, ^{14}C, and ^{222}Rn;
  - neutron logging to detect changes in water content and gamma-spectral logging to detect changes in radionuclide composition using access tubes installed to 5 m on a regular schedule and within 24 hr of precipitation events ≥2.5 cm.

The monitoring system (Figure 6.1) will be restricted to the upper 6 m to avoid penetrating the compacted amended clay layer or capillary barrier. Trench-cover monitoring will be conducted primarily adjacent to the monitoring islands. In addition, measurements of surface flux, plant transpiration, and vegetation sampling will be conducted at one randomly selected location within four randomly selected 15 m² sampling grids for each 60 m section.

Monitoring Beneath the Trenches

Monitoring beneath the trench constitutes downgradient monitoring in terms of the unsaturated zone. The objectives of monitoring beneath the trenches are to monitor changes in water content, and radionuclides in the gas and liquid phases. Neutron and gamma-spectral logging will be conducted in vertical access tubes installed to 30 m beneath the center of the bottom and sidewalls of each trench and also in slant access tubes installed beneath the sidewalls and base of each trench to a maximum depth of 3 m below the trench floor. These slant boreholes will be installed at one location between the monitoring islands. Soil-gas samples will be collected at depths of 1.5, 3, 6, 12, and 18 m beneath the base of the waste trenches in the center and sidewalls. Additional slant boreholes will be drilled to collect soil samples if other monitoring indicates significant migration of radionuclides.

Monitoring Adjacent to the Trenches

The proposed monitoring system adjacent to the trenches is similar to that within the trenches in evaluation of atmospheric releases of volatile radionuclides and quantification of concentrations and concentration gradients of volatile radionuclides in the trench-cover materials. In addition, soil-water balance of native sediments will be examined. The instrumentation used is similar to that installed in the trenches and penetrates to a maximum depth of 30 m below land surface. The culvert design is not required in this setting as the system does not penetrate the waste.

Trench-Cap Demonstration Unit

The Trench-Cap Demonstration Unit (TCDU) will be located outside the control area but within the fenced portion of the site. The dimensions of the TCDU will be approximately 46 m long and 30 m wide and 6 m deep. The TCDU will be divided into two segments to allow evaluation of both the A and B/C trench covers. Nonradioactive packages simulating the disposed waste will be emplaced in the TCDU. Instrumentation including neutron-probe access tubes, heat-dissipation probes (HDPs) and thermocouple psychrometers (TCPs) will be installed in the thickest section of the cover and in the native soil along the interface between the cover and excavation sidewall. The HDPs will be installed in a small bag of silica flour that is prewetted to a matric potential slightly higher than that of the soil. TCPs will be installed in 2.5 cm PVC casings that are slotted at the depth that the TCP is to be placed. A 0.63 cm air line will extend from the TCP casing to the surface and will be used to purge the system. Neutron-probe access tubes will consist of 5 cm aluminum casing. The HDPs and TCPs will be connected to an integrated data acquisition system for automated monitoring at hourly or daily frequencies. Monitoring of neutron access tubes will be event based.

Committee Observations on Proposed Performance
Monitoring of the Unsaturated Zone

-Major Issues-

The committee concludes that the foregoing description of the monitoring program demonstrates that there are detailed plans to conduct performance monitoring

during operation of the facility. Section 5300.1.1 of the license application states that the operational environmental monitoring program will continue "in full force and effect" for approximately 5 yr after closure of the facility; however, the environmental monitoring report (Atlan-Tec Inc., 1991) indicates that the sampling, analysis, and periodicity of the post-closure monitoring may be altered depending on analysis of the 30-yr operational monitoring. This post-closure monitoring will be conducted beneath the trenches which constitutes downgradient monitoring in terms of the unsaturated zone. In what follows, the committee offers some observations on the monitoring program and some suggestions to enhance the effectiveness of the program.

Unsaturated-Zone Monitoring Investigation and Action Levels

Regulatory investigation or action levels relative to the unsaturated-zone performance monitoring are not described in the revised environmental monitoring report (Atlan-Tec Inc., 1991). In section 4400.2.1.2.4 of the license application, however, action levels for unsaturated-zone monitoring consist of three standard deviations above average background levels or a continuous upward trend in concentrations. License condition 121 requires investigation levels for performance monitoring and license condition 122 requires reporting of time-series trend analysis to evaluate monitoring results (DHS, 1993).

Despite these various remarks on investigation and action levels relative to performance monitoring, the revised environmental monitoring program contains no discussion of these levels (Atlan-Tec Inc., 1991). The measured levels of contaminants in the unsaturated zone at which investigation and/or remedial actions should be initiated during performance monitoring have not been addressed in regulations governing LLRW disposal. This may be because regulations for LLRW disposal were developed primarily in relation to disposal in humid sites which have shallow water tables and thin unsaturated zones. Therefore, opportunities for conducting performance monitoring in the unsaturated zone in humid sites are limited. In contrast to humid sites, arid sites have thick unsaturated zones and allow detailed monitoring of concentration gradients to evaluate movement of radionuclides.

The committee strongly recommends developing investigation and action levels for performance monitoring in the unsaturated zone. In the unlikely event of contaminant movement through the unsaturated zone, it is not prudent to wait for contaminants to reach the compliance boundary before investigating the contaminant movement and developing an action plan. The Nuclear Regulatory Commission should provide guidance to low-level waste disposal facilities sited in arid regions on development of the levels necessary to initiate investigation or actions to remediate. Remediation plans should also be established if the action levels for performance monitoring are exceeded. **The committee is concerned that no documentation of remediation plans are in place in case problems arise such as leakage beneath the trenches. This is a critical issue that must be addressed.**

Integration of Monitoring and Performance Assessment

The committee has noted that the applicant's monitoring program does not specifically address the relationship between monitoring and performance assessment. The committee regards integration of monitoring and performance assessment as an important step that would greatly enhance both programs and ensure that the monitoring data being collected are reasonable. Continual updates of performance assessment models at least every 5 yr would act as a check on the monitoring data and would lead to accurate post-closure performance assessments.

Extension of Site Characterization Database during Operational Monitoring and Integration with Performance Assessment

As was mentioned in Chapter 3, the 1-yr monitoring data set collected prior to license application is inadequate to evaluate with a high degree of confidence subsurface flow in desert sites that are subjected to large interannual variations in rainfall. The proposed monitoring program will extend this database to a minimum of 35 yr. Although site characterization is not described explicitly in the proposed monitoring program, the proposed monitoring of ambient conditions is similar to site characterization but less comprehensive in scope. Long-term monitoring will also be integrated with performance assessment although the monitoring program does not specifically address the relationship between monitoring and performance assessment.

The integrated approach of site characterization, monitoring, and performance assessment modeling is not discussed in detail in the regulations; however, the committee supports this iterative approach that is now proposed by agencies such as the US NRC and DOE. It may result in expansion of the monitoring program but the committee considers it a reasonable and prudent monitoring philosophy. Periodic updates of performance assessment models, at least every 5 yr, will act as a check on the monitoring data and will permit taking advantage of newly developed performance assessment methodologies. Site characterization data collected during operations would supplement the fundamental site characterization data collected prior to the license application and improve the understanding of the unsaturated zone at Ward Valley.

Oversight Advisory Panel

Unsaturated-zone hydrology is a relatively young science, and new technologies are continually being developed. It will be critical to use state of the art technology when the monitoring program is implemented. **In view of the complexities involved in monitoring unsaturated zones in arid systems and the poor quality (discussed in Chapter 3) of the monitoring data collected for the license application, the committee recommends that future monitoring be directed and overseen by a panel of experts, some of whom would** be especially knowledgeable in the state-of-the-art of unsaturated-zone hydrology, soil physics, desert water-balance modeling, and ground-water hydrology in arid regions. These experts would also assist in reviewing the monitoring data and in recommending ways of rectifying any problems that may arise. In addition, the peer-review advisory panel would be involved in evaluating iterative processes of site characterization, monitoring, and performance assessment. Continuing scientific peer review would build credibility and public confidence in the monitoring program.

Other Issues and Recommendations

Long-Term Monitoring

The monitoring program includes a variety of techniques to monitor liquid and gaseous radionuclides. It is the committee's view that the monitoring instruments should be installed as early as possible to provide a long-term record of background levels. The length of time required for the proposed monitoring is much greater than has ever been conducted in the past. Great care and attention will be required to assure continuity of data collection and longevity of equipment, cross calibration between old and new instruments, etc. For example, the proposed program includes aluminum access tubes for neutron and gamma-spectral logging. Such access tubes corrode over relatively short time periods (years), and steel or some other longer lasting material should be used instead.

Technology Development

The proposed monitoring program includes a variety of technologies that provide multiple lines of evidence to evaluate performance of the facility. The committee thinks that some of the proposed measurements, such as porometer and flux-chamber measurements, are of questionable value; however, it is critical that some measurements for estimating evapotranspiration be made. The monitoring program was developed in 1991, and technology for monitoring subsurface water movement has improved since then. The monitoring program should be evaluated prior to implementation to incorporate technology developments. One of the limitations of the proposed plan is that it does not include automated monitoring of water content in the trench covers. Although event-based logging of neutron probes is proposed, this monitoring is limited to 24 hr after significant rainfall events ($\geq$ 2.5 cm). Since 1991, time-domain reflectometry has advanced considerably and is being implemented in many field sites and trench cap studies to monitor water contents in the soil. The committee recommends that time-domain reflectometry be used in the trench covers and in the TCDU.

Soil Air Pressure Monitoring

Monitoring migration of subsurface gaseous radionuclides such as ^3H, ^{14}C, and ^{222}Rn is an important part of the proposed monitoring program, but the issue of advective gas flow as a result of barometric pressure changes is not addressed. Monitoring of soil-air pressures could be readily incorporated into the proposed program because the gas piezometers will be installed for gas sampling.

Monitoring Ambient Conditions

The proposed monitoring program suggests that monitoring of the ambient conditions will be conducted immediately adjacent to the trenches, which means that the instrumentation will penetrate 0.6 to 0.9 m of trench-cover material. To evaluate true ambient conditions, additional instrumentation should be installed at a location distant from the engineered system. This monitoring of ambient conditions would act as a control site to provide background data against which monitoring data from the engineered facility can be compared.

Important information on site characteristics that should be collected during operations includes representative data on subsurface temperatures and water potentials that more accurately reflect ambient conditions and would help to avoid problems with ambient air temperatures on data loggers, as were found during prelicense monitoring. Because the trenches for the waste will extend to a depth of 18 m, it will be important to measure water potentials at depths ≥30 m because data from the Nevada Test Site revealed high water potentials in this zone that plot to the right of the equilibrium line of Figure 3.2, indicating downward drainage (Estrella et al., 1993). More information on spatial variability in subsurface flow processes should be collected, such as data to evaluate subsurface flow beneath Homer Wash and the borrow pit. This information will be valuable in conducting the performance assessment at closure.

Integration of Facility Design and Subsurface Monitoring

The monitoring plan does not discuss the facility design. **It appears to the committee that the facility design and the monitoring plan were developed independently; however, in the committee's judgement, integration of design and monitoring is critical for successful implementation of the monitoring program.** For example, drilling activities, either for routine monitoring or corrective action, must be considered in the cell design and spacing to make sure that ample room has been provided for slant drilling below the trenches. Monitoring systems must also be designed such that they do not compromise other systems. For example, monitoring of the trench covers must not interfere with the integrity of covers to water infiltration, yet sensors may be placed within the covers to monitor performance. The concept of monitoring islands is new and has not been evaluated at any facility. **Because the monitoring islands do not contain waste within or immediately surrounding the islands, it is likely that they will form local topographic highs and result in topographic relief on the surface of the cover enhancing any relief that could result from compaction of the waste. This topography would promote ponding and possibly create preferential flow pathways. This underlines the importance of evaluating differential settlement, which is not addressed in the proposed monitoring program.**

Design of engineered structures such as flood control berms that penetrate 3 m under the surface is discussed in the next chapter on flood control devices.

Trench-Cap Demonstration Unit

Studies proposed for the TCDU appear to be limited to consideration of A and B/C trench covers. The excavation depth of the proposed TCDU is 6 m. Critical layering in the B/C trenches is much greater than 6 m. **The committee recommends that the scale of the TCDU more closely match that of the actual facility; otherwise, the data may not be useful. Lysimeters should be used in these studies to provide more detailed information on water-balance issues. Vegetated and nonvegetated conditions should be used to simulate the effect of severe drought on subsurface flow. In addition, irrigated and nonirrigated conditions should be incorporated to evaluate possible effects of climate change including long-term climate change or several years of drought or above-normal precipitation. Different trench-cap designs could also be studied to evaluate their relative performance.** The TCDU will be instrumented with heat dissipation units and thermocouple psychrometers. Monitoring at the site with these instruments conducted prior to the license application was not very successful. **Because these instruments are not robust, they should be retrievable. A scheme for installing retrievable thermocouple psychrometers has been devised at the Beatty site (Prudic, 1994) and should be evaluated prior to this work. Because many of the technologies proposed for the TCDU are not straightforward, research should be conducted to ensure that reliable data will be collected.**

SUMMARY AND CONCLUSIONS

Monitoring in the Unsaturated Zone

• **With regard to the Wilshire group's comments concerning the absence of downgradient monitoring plans, it appears their concerns are not borne out by the administrative record. Definite plans are in place for post-closure monitoring downgradient in the unsaturated zone beneath the trenches. However, no remediation plans are described in the revised plan for the unsaturated zone.**

• Investigation and action levels for radionuclides relative to performance monitoring should be documented in the revised monitoring program and remediation plans should be documented if these action levels are exceeded.

• Monitoring and performance assessment should be integrated with continued site characterization. More emphasis should be placed on site characterization in the proposed monitoring program because ultimately reliance is placed on the natural system as the primary barrier to contain the waste. The 1-yr prelicense monitoring is insufficient in an arid system. Integration of monitoring and performance assessment would ensure that critical data are collected and would optimize sampling locations and frequency. Periodic updating of the performance-assessment models would provide a check on the monitoring data and would take advantage of improved performance-assessment methodologies.

• The poor quality of monitoring data collected for site characterization and the complexities of using state-of-the-art instrumentation during performance monitoring, in the committee's judgement, necessitates the establishment of an oversight committee of scientific experts to review the monitoring program and to assist DHS in assessing the data periodically.

• The monitoring program and facility design should be integrated to ensure that the design allows the proposed monitoring to be implemented and also to make certain that any engineered systems that could enhance infiltration are monitored.

• Because performance data for engineered barriers in arid settings are limited, significant emphasis must be given to design and performance testing of the TCDU and the scope of the TCDU studies should be greatly expanded to include all aspects of site characterization, monitoring, and performance assessment.

THE COMPLIANCE MONITORING PLAN FOR THE SATURATED ZONE

The Wilshire Group Position

The Wilshire group stated that in the license application, no plans were revealed for monitoring ground water downgradient from the site (Wilshire et al., 1993a). In their expanded discussion of earth science concerns (Wilshire et al., 1993b), they stated:

"Considering the complexity of the alluvial materials at the site, it would seem that off-site downgradient monitoring is an appropriate safeguard for which no provision has been made. This safeguard is desirable because escape pathways may exist at any point along the unlined trenches and the proposed monitoring system contains many gaps.... Moreover, the possibility of reversal of the gradient by pumping north of the site.... indicates that not every potential pathway was analyzed, as stated. Monitoring of off-site locations should be conducted by an independent party."

The DHS/U.S. Ecology Position

The Department of Health Services believes that given the extensive unsaturated-zone monitoring program, the existing monitoring wells MW-01 and MW-02 on the edge of the buffer zone (Figure 6.2), the low flux and long travel time in the unsaturated zone, and the low ground-water velocity in the alluvial aquifer, additional downgradient monitoring wells are unnecessary (Brandt, 1994). The DHS believes that there is no justification for additional monitoring wells to the south of the 1000 acres to be controlled by the State. It also argues that there is no justification for any additional monitoring wells to the north of the disposal site to protect against projected drawdowns in hypothetical wells that might possibly be developed at some future time in lower Piute Valley or the northern portion of Ward Valley. DHS believes it is more reasonable to implement monitoring of ground-water levels if and when such well field is developed to determine the effect it might have.

The Committee's Approach

In the view of the committee, monitoring of the saturated zone is a secondary defense, and the most comprehensive monitoring should take place in the unsaturated zone. Nevertheless, saturated-zone monitoring should be designed to be comprehensive and useful both for background information and for interpreting possible impacts caused by migration of radionuclides to the saturated zone. The fluid-potential data at the site suggest a southerly gradient in the saturated zone.

Investigation and Action Levels

U.S. Ecology summarized the proposed saturated-zone monitoring decision and response process in a decision tree and accompanying text (U.S. Ecology, 1991, Fig. IV.B.a). The operational period monitoring program is based on "action-level" and "investigation-level" criteria. If an investigation level criterion is exceeded, then a series of evaluations and further investigative steps are triggered to assess the significance of the criterion exceedance. If an action level criterion is exceeded, then another series of investigative and evaluative steps are triggered. If the exceedance is confirmed, then the cause of the exceedance is to be determined and an appropriate response proposed to the DHS. After receiving DHS approval of the proposed response, the response would be implemented.

Detailed specific response actions that might be implemented if ground-water monitoring data exceed action levels are not presented by the applicant, although they do provide a listing of some of the types of actions that might be taken. These range from more dense spatial and temporal data collection to confirm and delineate the area of exceedance, to exhuming the waste and/or ground-water remediation. If an action level is exceeded, then whatever response action is taken must be approved by the regulator (DHS).

164

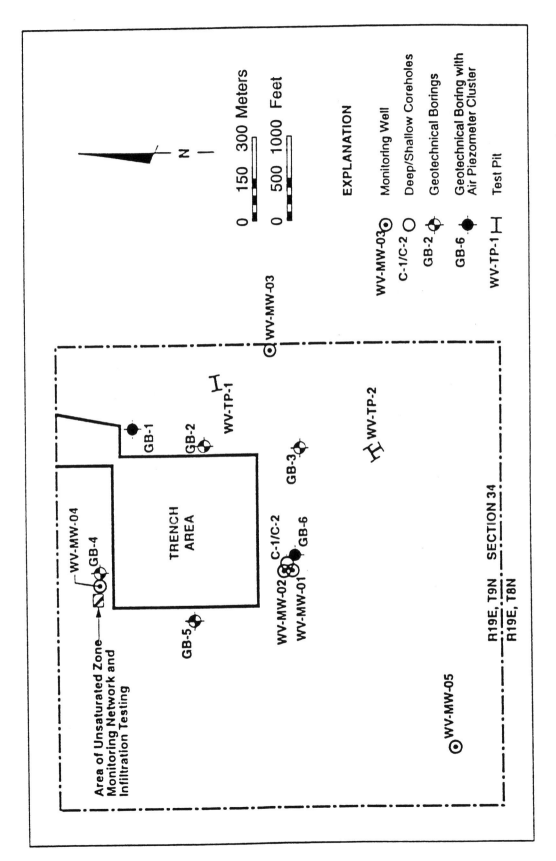

Figure 6.2   Details of Section 34 with location of site, wells, and borings (Harding Lawson Associates, 1994)

Ground-Water Monitoring

It is stated in the license application that long-term operational ground-water monitoring will be limited to four locations. One existing well, MW-04, will be upgradient from the facility. Three downgradient wells are proposed (wells MW-10, 11, 12). These locations are shown in Figure 6.2. In the latest information the committee received from U.S. Ecology, it is stated that monitoring will likely continue in existing monitoring wells MW-01,02, 03, 05 (Harding Lawson Associates, 1994). The new monitoring wells will be screened in the upper 15 m of the saturated zone and the lower 3 m of the unsaturated zone. Wells will be sampled quarterly during the operational and post-closure period. Ground-water samples will be analyzed for major ions and basic radiological parameters (LA Section 4400, Table 4400-1). A parameter value larger than the background value plus three standard deviations will be considered a statistically significant change in solute concentration. The limited number of monitoring wells in the saturated zone is defended by U.S. Ecology on the grounds that the comprehensive unsaturated-zone monitoring program ensures site performance (Harding Lawson Associates, 1994).

Committee Observations and Recommendations on Ground-Water Monitoring

Investigation and Action Level Response Plan

It is the committee's view that the response plan to investigation or action-level exceedance, although lacking details at this stage, generally appears to be adequate and appropriate. Without knowing the details of the nature of the action-level exceedance and its causes, it is not feasible to state in advance what specific response actions would be most appropriate. However, the plan appropriately places responsibility on the operator to develop an adequate response plan that is acceptable to the regulator. **The process outlined to assess the exceedance data and to determine the proper response action appears to the committee to be adequate.**

Oversight Advisory

The committee reiterates here the recommendation that the applicant consider an independent scientific oversight advisory panel of experts to review and make recommendations on all monitoring data. In the event of an exceedance of an investigation-level or action-level criterion, the oversight panel committee would review the data and the proposed response plan and present recommendations to the DHS.

Placement of Monitoring Wells

The committee considers that the proposed spacing of monitoring points along the southern and eastern perimeters of the radiological control area may not be adequate (Figure 6.2), given the lateral extent of the facility and the likely areal extent of any contaminant plume at the water table. It is our opinion that each of the southern and eastern perimeters of the radiological control area has no fewer than four monitoring wells, inclusive of corner monitoring locations (i.e. a total of eight monitoring wells). In addition, to establish better background databases, the western and northern perimeters should be equipped with no fewer than three monitoring points (i.e. a total of three background wells).

In the view of the committee, although the site-characterization borehole MW-02 is not suitable for monitoring the water table, it is suitable for monitoring deeper migration of contaminents.

The committee does not see the need to position any ground-water monitoring wells beyond the boundaries of the disposal facility.

Well Construction

We recommend that the monitoring wells be of appropriate construction and design to ensure monitoring of the uppermost saturation zone, (e.g., screened interval from 1.5 m above to 3.5 m below the water table), and a deeper zone (arbitrarily an interval 28 to 33 m below the water table). In this manner, water chemistry changes from samples at two horizons at the same location, as well as fluid potential within the two horizons, will be established. These databases are more useful to recognize changes occurring in water chemistry over time caused by leachate reaching the zone of uppermost saturation. Furthermore, they may detect changes in vertical fluid potential that could be related to local recharge.

Drilling Methodology

We recommend that the methodology used for drilling the monitoring boreholes be based on a dual-wall reverse-circulation air-rotary system. This technology produces the best control during drilling for identifying the depth of apparent uppermost saturation and permeable zones, as well as vertical differences in fluid potential. The technique creates a clean borehole for final well configuration and also tends not to damage significantly formation properties next to the borehole. The committee believes this basic drilling technology and strategy of vertical-profile monitoring, beginning at the uppermost saturation and extending to the first zone of important relative permeability as indicated by air-lifted water production during drilling, would be appropriate for the primary objectives of the saturated-zone monitoring.

REFERENCES

Atlan-Tec Inc. 1991. Environmental Monitoring Report for the Proposed Ward Valley Low-Level Radioactive Waste Disposal Facility. Prepared for the Department of Health Services.

Brandt, E. C. October 7, 1994. Summary submittal of the California Department of Health Services to National Academy of Sciences Committee to review specific scientific and technical issues related to the Ward Valley, California LLRW site.

Campbell, A. C., F. W. Ross, and T. J. Nicholson. 1993. Low-level radioactive waste performance assessment technical issues and branch technical position. Proceedings of the Symposium on Waste Management. Tucson, Arizona. 367-370 pp.

Department of Health Services. 1993. Radioactive Material License issued to U.S. Ecology. September.

Estrella, R., S. Tyler, J. Chapman, and M. Miller. 1993. Area 5 site characterization project-report of hydraulic property analysis through August 1993. Desert Research Institute, Water Resources Center. 45121:1-51.

Harding Lawson Associates. 1994. Letter Report to Ms. Ina Alterman; Interpretation of groundwater quality data; Ward Valley. California. Dated October 6, 1994.

License Application. 1989. U.S. Ecology, Inc. Administrative Record, Ward Valley Low-Level Radioactive Waste Disposal Facility, Section 4400.

Prudic, D. 1994. Personal Communication.

U.S. Code of Federal Regulations. 1982. Licensing Requirements for Land Disposal of Radioactive Waste. Part 61: Title 10. Washington, D.C.: U.S. Government Printing Office.

U.S. Ecology. 1991. Environmental Monitoring Report for the Proposed Ward Valley Low-Level Radioactive Waste Disposal Facility. Prepared for the Department of Health Services. December 12.

Wilshire, H. G., K. A. Howard, and D. M. Miller. 1993a. Memorandum to Secretary Bruce Babbitt. June 2.

Wilshire, H. G., K. A. Howard, and D. M. Miller. 1993b. Description of earth science concerns regarding the Ward Valley low level radioactive waste site plan and evaluation. Released December 8.

7

FLOOD CONTROL AND ENGINEERING CONSIDERATIONS

Issue 5
The Potential for Flood and Erosion Control Devices to Fail

THE WILSHIRE GROUP POSITION

The Wilshire group has expressed concern about the long-term stability of engineered flood and erosion control facilities. In their first memorandum (Wilshire et al., 1993a), referencing the draft EIR/S, they referred to possible failure of "flood control devices" without specifying which devices they were addressing. In their expanded report (Wilshire et al., 1993b), they referred to "site evaluation documents that incorrectly claim" that the upslope diversion berms will protect the primary flood control berm against erosion. In addition, they predicted channelization of surface water at the ends of the diversion berms and consequent threats to the integrity of the trench cover. Thus their two main concerns are: (1) the proposed flow diversion or breakup berms upslope to the west of the 70-acre LLRW site and (2) the proposed rip-rapped flood-protection barrier around the LLRW site. The Wilshire group questioned the purpose or function of the breakup berms and postulated that channelization at the end of the failed berms would force concentrated runoff toward the western edge of the main rip-rapped flood-protection barrier. They claimed that water would pond in front of the barrier, giving rise to potential leakage into the trenches. They raise an additional concern that channelized runoff from the breached breakup berms would undercut the rip-rapped flood-protection barrier and cause slope failure (Wilshire et al., 1993b; Wilshire et al., 1994).

THE DHS/U.S. ECOLOGY POSITION

DHS has indicated that the design of the durable flood-protection structures at the disposal site is based on conservative U.S. NRC guidance to protect LLRW facilities. This guidance is based on Corps of Engineers design procedures for erosion protection during large storms on alluvial fans, and on U.S. NRC guidance for erosion protection at uranium mill tailings facilities. DHS reputes that this guidance "represents the state of the art in hydraulic design and erosion protection" (Brandt, 1994).

THE COMMITTEE'S APPROACH

The following discussion about the erosion-control and flood-control facility features is based on the committee's examination and analyses of (1) available design documentation, including appropriate license application materials; (2) summary reports and materials provided by both opponents and proponents of the project; (3) written information provided during the two NAS/NRC committee meetings in Needles in July and August/September, 1994; (4) field visits to the facility site and to nearby I-40 drainage facilities in July and August, 1994; and (5) other relevant scientific references.

DESCRIPTION OF PROPOSED FACILITIES

Location and general description of the Ward Valley low-level radioactive waste (LLRW) disposal facility has been presented in Chapter 2. A detailed description of the major elements of the proposed LLRW facility which involve engineered flood and erosion control facilities follows.

Radiological Control Area Facilities

General Trench Area Design

The proposed Radiological Control Area consists of a 532^{1} m by 532 m (approximately 28 hectares) area surrounded by a 0.9 m to 1.5 m high flood protection berm and an electrified security fence, within which near surface LLRW disposal operations will take place in a series of five progressively developed trenches. Four unlined Class A waste trenches and one unlined Class B/C trench are planned and are illustrated in Figures 7.1, 7.2, and 7.3 (U.S. Ecology, 1990).

A planned buffer zone will extend 122 m around the perimeter of the fenced control area for carrying out environmental monitoring activities, maneuvering construction equipment, and allowing corrective actions to be implemented.

The base of the Class A trenches will be 18 m below the natural surface, and the Class B/C trench will be 13 m deep to eliminate any potential lateral flow from the bottoms of the Class A trenches into the B/C trench. All waste will be buried up to 6.1 m below the original ground surface and beneath the deepest projected scouring action predicted by a Probable Maximum Flood (PMF).

The four Class A trenches are about 470 m long by 88 m wide at final grade and 34 m wide at the bottom. The interior side slopes of these trenches are a ratio of 1.5 horizontal (H) to 1 vertical (V) (1.5H:1V). A 12 m wide space at the surface separates each completed

---

[1] All conversions from English system (feet/pounds) to metric (meters/grams) have been rounded to two significant figures. See conversion table in Appendix C.

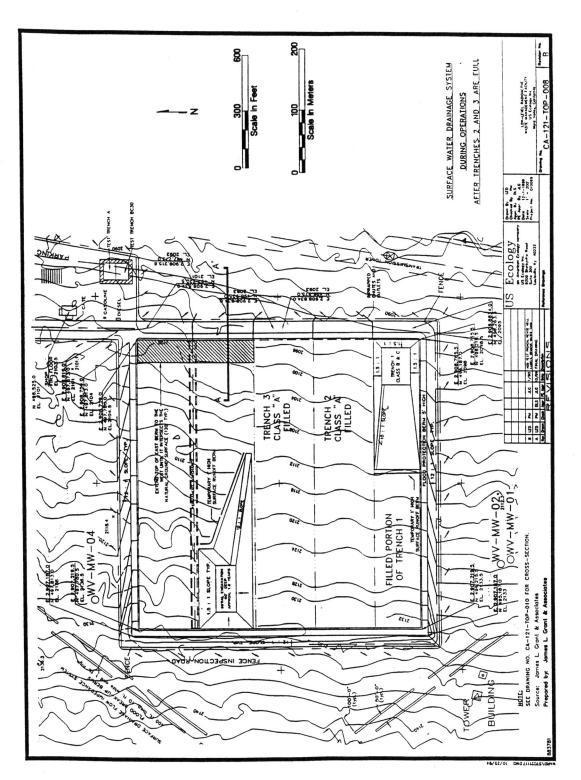

Figure 7.1 LLRW control site, perimeter flood-protection berm, part of the upslope breakup berms, typical excavations for both trench classes, and other site features superimposed on an existing topographic contour map for the area. (James L. Grant & Associates, 1989)

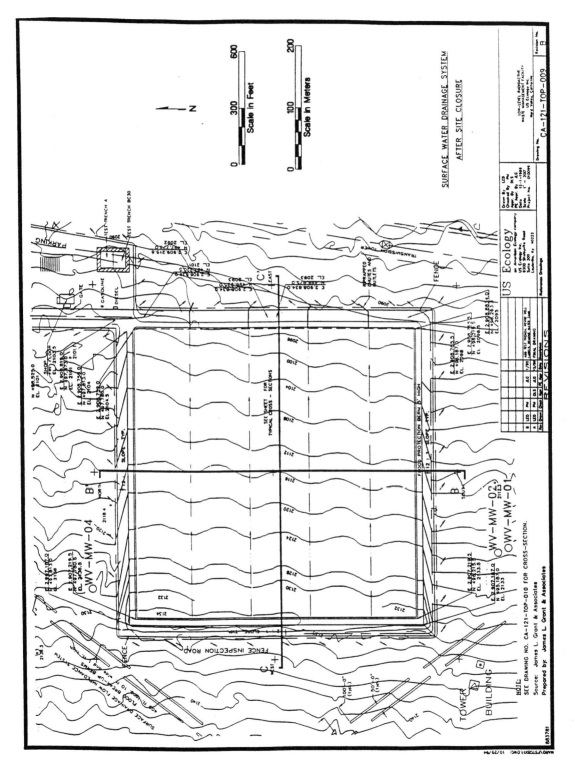

Figure 7.2 LLRW disposal site showing finished grade contours and cross section locations through trench (James L. Grant & Associates, 1989)

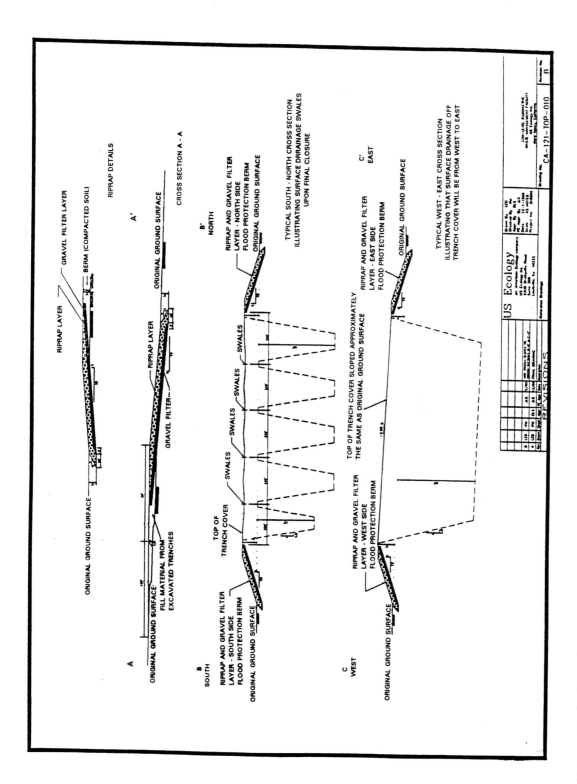

Figure 7.3 Typical south-north (BB') and west-east (CC') cross sections through trench and other typical sections (James L. Grant & Associates, 1989) (Not converted to metric scale; see Appendix C for conversion factors).

trench. The Class B/C trench is about 470 m long by 69 m wide at final grade and 31 m wide at the bottom. The inside side slopes of this trench are also 1.5H:1V. All five trenches will be excavated parallel to the slope of the ground surface and the approximate dip of sediments comprising the alluvial fans. Both Class A and B/C trench bottoms will slope about 2 percent west to east away from the direction of disposed wastes.

The trench covers for all five disposal trenches will consist of a 2.4-m thick silty sand vegetative-support layer overlying 5.2 m of reworked sediments that were removed from the trenches. This means that the total trench cover thickness of 7.6 m includes the 6.1 m cover of fill over both classes of waste plus a permanent 1.5-m high flood-protection cap/berm.

The waste disposal trench cover is designed to shed surface flow resulting from the Probable Maximum Precipitation (PMP) storm event, and the resulting surface runoff will be directed laterally into four shallow 0.3 m deep swales which slope about 2 percent toward the east. Drainage from the swales and off the cover will be directed across a system of rip-rapped chutes and outlet controls along the edge of the eastern berm toward Homer Wash. No design details are apparent in the drawings or license application for the chutes and outlet controls.

Onsite storm water that falls outside the trench, but inside the perimeter flood-protection berm during active operation, will be prevented from entering the trenches by a planned combination of temporary berms and ditch system to divert water from the excavated trench areas over the natural ground.

No specific management plan is indicated for dealing with rain water that falls directly into the open trenches and collects during active waste disposal operations, other than through natural evaporation, removal if necessary, or "proper" handling (LA Sections 3100.1.1 and 3200.4). The California DHS noted in its Summary Report, however, that "standing water or moist sediments will be removed from the trenches and tested for radioactivity" (Brandt, 1994).

Flood Protection Berm

The design plans for the site indicate that offsite storm water will be prevented from entering the trench area during operation and after closure by a permanent flood-protection berm surrounding the disposal site. The berm, which is to be constructed when site construction begins and incorporated into the final site cover at closure, is designed to withstand the Probable Maximum Flood (PMF). The berm will rise 1.5 m above the original ground surface on the upslope (western) side to 0.9 m on the downslope (eastern) side and is designed to prevent overtopping by wind and wave action during the PMF event. The purpose of this berm is to contain surface-water flow during a PMF event and, from the approximately 10 km^2 drainage area upslope from the facility, for the same flood, to divert flow around the north and south sides of the facility during operations and after closure. The permanent berm has 12H:1V outer exposed side slopes and 1.5H:1V interior side slopes.

Embankment armoring consisting of a 0.9 m thick layer of 0.6 m average size stone rip rap and a 45.7 cm thick filter base of 7.6 cm maximum size gravel is proposed to stabilize the surface against wind and water erosion. The outer embankment armoring system is to be extended at a 2H:1V slope to a depth of 1.5 m into the subsurface to provide some scouring protection from adjacent surface water flow and to interrupt any lateral flow of infiltrating precipitation through the shallow calcrete layers.

The outer stone flood protection berm represents about 3 ha (7 acres) of exposed surface area, or about 10 percent of the approximately 28 ha (70-acre) radiological control area. This highly permeable surface will encourage directly-falling precipitation, or non-flood water, to infiltrate into the berm slope fill and possibly add recharge water to the edges of the trench zone over several decades of exposure.

Because the upstream toe of the facility flood berm is nearly parallel with the topographic contours, little gradient exists for water to flow away from that side, and there could be a tendency for water to pond in the middle of the western side. No floodwater, however, is expected to encroach on the disposal site from Homer Wash, which is located about 760 m to the east, because the nearest natural surface elevation at the site is reported to be over 13 m above the estimated 100-year flood level and 13 m above the estimated PMF elevation. Interpretation of these data indicates that the bottom of the Class B/C trench would be about 0.6 m higher than the Homer Wash PMF peak and 1.5 m *higher* than the 100-year flood, but the Class A trenches would be about 4 m and 5 m *lower* than these floods, respectively.

Breakup Berms

Offsite storm water flows onto the LLRW site from the west as sheet flow or in small rills. A series of shallow-flow breakup berms will be placed in a staggered, offset chevron pattern upslope and west of the disposal facility, (Figure 7.4), primarily (1) to increase sheet flow roughness for the purpose of reducing the sheet flow velocity and inducing more tranquil subcritical hydraulic conditions near the permanent primary flood control berm to reduce scour potential and (2) to divert storm runoff to the north and south of the facility (LA Section 3310.2). Each berm that forms part of the chevron pattern would be 122 m long, 3 m wide and 0.3 m high. These shallow flow breakup berms, however, are intended to be temporary features that will be constructed during the initial trench excavation with materials removed from the trenches and maintained during the operations and institutional control periods. A more detailed analysis of the breakup berms is discussed later in this chapter.

HYDROLOGICAL SETTING

The general Ward Valley Watershed geology, hydrology and climatology were described in Chapter 2. Details of the hydrological characteristics of the 127 km^2 Homer Wash Watershed upstream from the LLRW site and the 9.8 km^2 local site subbasin hydrology

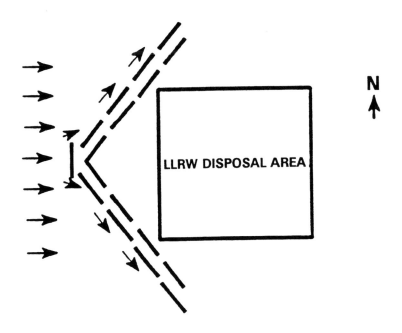

Figure 7.4   Schematic plan view of temporary breakup berms.

which drains toward the LLRW from the west are described in Box 7.1 and Box 7.2, respectively. These summarize the information from which the committee has drawn certain conclusions about the ability of the design to divert water from the radiological control area and about the stability of the flood-control berm. The hydrologic and hydraulic information were extracted from the administrative record license application (sections 2410, 3200, 3200.2, 3440, and 5110) and from the license application interrogatories/responses (0449A3440.2 through 0453A3440.2, 0454A3440.4 thorough 0454A3440.4, and 0449B3440.2) and from the DHS summary report (Brandt, 1994).

Homer Wash Watershed

A complete hydrologic description and analysis of Homer Wash Watershed above the LLRW site was needed to define the proximity of the site to the estimated 100-year flood plain and the Probable Maximum Flood (PMF) (Box 7.1); to classify the FEMA flood zone for the site; and to determine the short- and long-term flood protection requirements, if necessary, against the PMF. Because no historical flow data were available for Homer Wash

Box 7.1

Design Storm Parameters for the Homer Wash Watershed

24-hour, 100-year storm depth...3.5 in (8.9 cm)[1]
1 sq. mile, 1 hr. Prob. Max. Precip. (PMP) depth.........................10.9 in (27.7 cm)
1 sq. mile, 6 hr. PMP depth ....14.7 in (37.3 cm)
49.2 sq. mile, 6 hr. PMP depth for Homer Wash.........................10.7 in (27.2 cm)

HEC-1 Computer Program Flood Peak Results
for Homer Wash:
24-hr, 100-year flood flow peak.. 1,461 cfs (41.4 m³/s)
6-hour PMF .... 19,230 cfs (544.6 m³/s)

Estimated LLRW Site Disposal Area Elevation Above Calculated Peak
Flood Level, using HEC-2 Computer Program Water-Surface Profile
Analysis:
100-year Flood...47 ft (14.3 m)
6-hour PMF ....44 ft (13.4 m)

[1] Conventionally, English units are used for this type of measurement

and its tributaries, appropriate hydrologic data were developed for conducting computer simulation modeling for various assumed floods. The following data summarize the Homer Wash Watershed characteristics upstream from the LLRW site:

Drainage Area	127.4 km²

Soil Conservation Service (SCS) Hydrologic Soil Group (described in Chapter 2):

Zone 1 (HSG "D")	32% of Drainage Area
Zone 2 (HSG "C")	0%
Zone 3 (HSG "B	13%
Zone 4 (HSG "A	55%

The Homer Wash Watershed was subdivided into 17 sub-basins. Standard SCS procedures were used to determine parameters such as SCS weighted Curve Number, which combines the effects of soil type, surface cover and land use, and assumed antecedent soil moisture condition (AMC) plus time of concentration and lag time (USDA/SCS, 1986; USDA/SCS, 1972). These parameters are summarized in Table 7.1 for each of the sub-basins (LA Section 2410.3.1).

The hydraulic lengths for the 17 sub-basins range from 1,676 m to 12,800 m; the average sub-basin watershed land slopes range from 1.5 to 14 percent; the SCS Curve Numbers vary from 52 to 80, and the times of concentration from 1.03 to 4.18 hours.

The studies indicate, based on the above hydrologic data and computer results, that the site is located well above the Homer Wash floodplain for floods through the 100-year flood and up to the Probable Maximum Flood (PMF), which is based on Probable Maximum Precipitation (PMP). The PMF is the flood that may be expected from the most severe combination of critical meteorologic and hydrologic conditions that are reasonably possible in the region. While the PMF is considered a rare event that is not generally associated with a fixed return period or exceedance probability, it is typically assigned a return period of 1,000 to 1 million years (Moser, 1985). A 100-year flood is an event that has a 1 percent chance of occurring or being exceeded in a given year (or once, *on average*, every 100 years). From the principles of probability, however, the chance of a 100-year flood occurring during any given 100-year interval is about 63 percent. To put the current LLRW site situation into perspective, the probability of a 100-year flood occurring during the 30 years of active facility operation is 26 percent. For the expected 500-year project design period, the chance of a 100-year flood happening is 99.34 percent. These probability calculations do not consider any future effects of global warming, land-use changes, or other long-term temporal changes that would affect precipitation or runoff processes.

The committee agrees with the license applicant that I-40, which acts like a low dam across Homer Wash, does not represent a realistic threat to the LLRW site. The likelihood is small that the road will fail and threaten the LLRW site. Dam failure analysis is, therefore, not necessary in the Homer Wash flood analysis.

Local Site Surface Hydrology

The local site watershed is defined as the area upgradient to the west of the site that could contribute storm-water runoff over the site. This area is given as sub-basin 65 (Table 7.1). The upslope local drainage rises to the west to a maximum elevation of about 1086 m above mean sea level (AMSL) in the Piute Mountains. This compares to an elevation of about 645 m in the middle of the facility. The estimated PMF and its associated Probable Maximum Precipitation (PMP) (Box 7.2) were chosen as the bases for the design of flood protection features for the facility during operations, as well as for the site closure and stabilization period (LA Section 3440.2). While the use of the PMF is clearly acceptable for the relatively short-term operational design of low-level waste facilities, its use is not required, and a less conservative design basis such as a

179

Drainage Basin	Total Area (sq mi)	Hydraulic Length (feet)	Average Basin Slope (%)	Zone 1 Area (sq mi)	Zone 1 Hydrologic Soil Group	Zone 1 Vegetative Type	Zone 1 Curve Number	Zone 3 Area (sq mi)	Zone 3 Hydrologic Soil Group	Zone 3 Vegetative Type	Zone 3 Curve Number	Zone 4 Area (sq mi)	Zone 4 Hydrologic Soil Group	Zone 4 Vegetative Type	Zone 4 Curve Number	Weighted Basin Curve Number	Time of Concentration (Hours)	Lag Time (Hours)
5	5.10	42000	1.46	0.00	D	NA	NA	5.10	B	Desert Shrub (poor)	77	0.00	A	NA	NA	77	7.26	4.36
10	0.93	9800	3.73	0.46	D	Desert Shrub (poor)	88	0.47	B	Desert Shrub (fair)	72	0.00	A	NA	NA	80	1.03	0.62
15	5.10	33700	6.93	1.84	D	Desert Shrub (poor)	88	0.61	B	Desert Shrub (fair)	72	2.65	A	Desert Shrub (fair)	55	69	2.43	1.46
20	0.49	11300	1.57	0.02	D	Desert Shrub (fair)	86	0.00	B	NA	NA	0.47	A	Desert Shrub (fair)	55	56	2.19	1.31
25	6.20	28400	6.49	1.92	D	Desert Shrub (poor)	88	0.00	B	NA	NA	4.28	A	Desert Shrub (fair)	55	65	2.28	1.37
30	1.90	22600	5.61	0.55	D	Desert Shrub (fair)	86	0.00	B	NA	NA	1.35	A	Desert Shrub (fair)	55	64	3.18	1.91
35	4.00	21300	5.87	0.96	D	Desert Shrub (poor)	88	0.00	B	NA	NA	3.04	A	Desert Shrub (fair)	55	63	1.78	1.07
40	2.40	21600	13.52	1.30	D	Desert Shrub (poor)	88	0.00	B	NA	NA	1.10	A	Desert Shrub (fair)	55	73	1.53	0.92
45	3.50	20600	9.37	1.40	D	Desert Shrub (poor)	88	0.00	B	NA	NA	2.10	A	Desert Shrub (good)	49	65	1.38	0.83
45A	0.29	5500	2.00	0.00	D	NA	NA	0.00	B	NA	NA	0.29	A	Desert Shrub (fair)	55	55	1.07	0.64
50	5.70	30600	5.61	4.10	D	Desert Shrub (poor)	88	0.00	B	NA	NA	1.60	A	Desert Shrub (good)	49	77	2.14	1.28
55	4.60	28000	10.75	0.97	D	Desert Shrub (poor)	88	0.00	B	NA	NA	3.63	A	Desert Shrub (good)	49	57	1.68	1.01
60	0.40	9300	1.90	0.04	D	Desert Shrub (fair)	86	0.00	B	NA	NA	0.36	A	Desert Shrub (good)	49	52	1.71	1.03
65	3.80	31000	6.71	1.14	D	Desert Shrub (poor)	88	0.00	B	NA	NA	2.66	A	Desert Shrub (good)	49	61	2.39	1.43
70	0.35	11600	2.37	0.03	D	Desert Shrub (fair)	86	0.00	B	NA	NA	0.32	A	Desert Shrub (good)	49	52	1.75	1.05
75	3.20	34800	2.81	0.83	D	Desert Shrub (poor)	88	0.00	B	NA	NA	2.37	A	Desert Shrub (good)	49	59	4.18	2.51
80	0.83	14200	1.97	0.07	D	Desert Shrub (poor)	88	0.00	B	NA	NA	0.76	A	Desert Shrub (good)	49	53	2.05	1.23

Table 7.1 Homer Wash sub-basin hydrologic summary (Note: Sub-basin 65 represents the local site drainage hydrology) (LA Section 2410) (Original data not converted to metric units; see Appendix C for conversion factors).

Box 7.2

Design Storm Parameters for the Local Site Watershed

24-hour,100-year storm depth...3.5 in   (8.8 cm)[1]
1 sq. mile, 1 hr. Prob. Max. Precip. (PMP) depth .... 10.9 in   (27.7 cm)
1 sq. mile, 6 hr. PMP depth..14.7 in   (37.3 cm)
4.0 sq. mile, 6 hr. PMP depth for local drainage ....14.0 in   (35.6 cm)

HEC-1 Computer Program Flood Peak Results
for the Local Draining Area:

24-hour, 100-year flood flow peak   223 cfs (6.3 m^3/s)
6-hour PMF   10,270 cfs (290 m^3/s)

Hydraulic calculations produce the following results:

Maximum PMF Stillwater Depth Along West Edge
  of Permanent Flood Protection Berm.......................1.25 ft   (0.381 m)(3 ft used)
Maximum PMF Velocity Along West Edge
  of Permanent Flood Protection Berm........................4.8 fps (1.5 m/s)
Maximum PMF Stillwater Depth Along South Edge
  of Permanent Flood Protection Berm........................1.5 ft (0.46 m)
Maximum PMF Velocity Along South Edge
  of Permanent Flood Protection Berm........................6.4 fps (2.0 m/s)
PMF Wave Runup and Wind Setup Adjacent
  to the West Edge of Flood Protection Berm.............2.0 ft (0.61 m)
Calculated Water Depth + (Wave Runup +
  Wind Setup) Height:

3.0 + 2.0 = 5.0 ft (berm height used) (1.5 m)

Maximum PMF Flow Velocity on the Disposal Site Cover ....4 fps (1.2 m/s)
Maximum PMF Flow Depth on the Disposal Site Cover.......................0.4 ft (0.12 m)
Range of Calculated Maximum PMF Scour Depth Adjacent to
  Flood Protection Berm   5.6 to 20 ft (20 ft used) (6.1 m)

[1] Conventionally, English units are used for this type of measurement

100-year flood event could be used. Here, the PMF peak flow is conservatively assumed for engineering the Ward Valley LLRW site long-term flood and erosion protection facilities, which include the rip-rapped flood protection berm around the waste site and the trench cover swale and drains. No clear hydrologic flood criteria or basis for designing the upslope temporary breakup berms are apparent in the license application reports, interrogatory, and response sections (LA Sections 3100, 3200, 3400) in the responsive summary to comments to the Final EIR/S, or in the other hydrologic or hydraulic analysis documents.

As was the case for Homer Wash, no historical flow data were available for the local site sub-basin, and appropriate hydrologic data were developed for conducting computer simulation modeling for various assumed floods. The following data summarize the reported local site watershed characteristics upstream from the LLRW site:

| Drainage Area: | 9.8 km^2 (3.8 mi^2) |
| | (10.3 km^2 used)* |

SCS Hydrologic Soil Group (described in Chapter 2):

Zone 1 (HSG "D")	30% of Drainage Area
Zone 2 (HSG "C")	0%
Zone 3 (HSG "B")	0%
Zone 4 (HSG "A")	70%

* 3.8 mi^2 was rounded to 4 mi^2 for calculations which converted to 10.3 km^2

The local drainage area upslope from the LLRW site is one of the 17 sub-basins of the 127 km^2 Homer Wash Watershed. This is an estimated drainage area based on conditions that existed during the hydrological assessment. Although variations in drainage area can occur over time, the committee believes that the number used is conservative enough to represent a reasonable degree of engineering certainty. Parameters such as SCS weighted Curve Number, which combines the effects of soil type (see Table 7.1), surface cover and land use, and assumed antecedent soil moisture condition (AMC), along with time of concentration, and lag time, were determined using the same standard SCS procedures as were used for the Homer Wash Watershed.

The reported hydraulic length for the long and narrow local sub-basin is 9,450 m, the average sub-basin watershed land slope is 6.7 percent; the SCS Curve Number is 61 (assuming an average Antecedent Moisture Condition, or AMCII); and the Time of Concentration is 2.4 hours.

The hydrologic analysis for the local site also employed the special case FEMA flood zone classification and computations for alluvial fans and bajadas (LA Section 2410.7; FEMA, 1985). **Applying the current FEMA methodology to the proposed site, computations showed that the site flood zone parameters are well within the FEMA criteria for flood peak and flow depth to classify the proposed LLRW site outside Zone A (100-year floodplain)[2]. The Committee concludes that flooding resulting from upstream drainage would not occur (LA Section 2410.7).**

DESIGN CRITERIA

In challenging the flood and erosional controls planned for the site, the Wilshire group in effect was challenging aspects of the design of the proposed facility. Therefore, the committee assessed the criteria and effectiveness of the proposed design of the flood and erosion control structures. The driving force behind a *minimum* acceptable design is a regulation or code standard; however, public safety and health, common sense, experience, and professional judgement should determine whether a given design situation merits the need to *exceed* the minimum requirement.

Permanent Flood Protection Berm

The PMF (including its PMP storm derivative) was chosen as the basis for designing the height of and the erosion control for the permanent flood protection berm which surrounds the LLRW site. While the governing regulations (U.S. NRC, 1988) allow a 100-year flood criterion, the more conservative PMF criterion is certainly acceptable and prudent for protection of this type of facility over the expected long-term life of this project. **In examining the calculations for determining the berm height, the committee observed that by using the PMF, the berm height exceeded the regulatory requirement by 30 percent. If one assumes that the PMF event has a return period of 100,000 years, the probability of this magnitude of flood occurring during the assumed 500-year facility lifetime is less than 0.5 percent; at a return period of 10,000 years, the probability is less than 5 percent.**

---

[2] This methodology has been in use for about a decade and remains the state of practice recommended by FEMA. The methodology is currently being reviewed by the National Research Council *Committee on Understanding Alluvial Fan Flooding,* but any recommendations developed by that committee will not be available for some time.

Scour Depth

Three different methods were used to estimate the maximum PMF scour depth along the protective flood berm. Examination of the computation details shows that, while the range of predicted scour depth was 1.7 m to 6.1 m, the more conservative 6.1 m has been used in the license application. Because all trenches include 6.1 m of soil backfill between buried waste and the *original ground surface* at the time of completion, all waste would be buried below this deepest projected scouring action by the PMF. Should 6.1 m depth of scour occur, it would take place along the *outside* toe of the 12H:1V sloped and rip-rapped flood protection berm, or approximately 18 m horizontally away from the top edge of the berm and nearly 31 m away from the nearest trench. Since the flood-protection berm stone rip rap would extend vertically downward another 1.5 m below natural grade at the external toe and outward at a 2H:1V slope, or 3 m more, any significant (>1.5 m) vertical scouring action in the vicinity of the external toe of the flood-protection berm, resulting from the PMF, would likely begin 34 m away from the nearest trench. If an extreme event were to result in a greater scour depth, remedial measures might be needed to repair damage to the berm.

Rip-Rap Construction

The proposed rip rap for the external flood protection berms is 0.6 m average diameter durable stone. No other rip-rap size distribution parameters were specified in the design description. For a specified unit weight of 2640 kg/m^3, the *average* rock weight would be almost 318 kg. The rip rap is to be 0.91 m thick and overlaid on a 46 cm gravel filter bed. The U. S. Bureau of Reclamation has found a 0.91 m thickness of dumped rip rap to be generally most satisfactory for its *major* dams (U.S. Bureau of Reclamation, 1987). The rip-rap size, thickness, and filter system represent a reasonable design for embankments of this small size and subject to the specified flow forces.

Members of the committee inspected two engineered training dikes for controlling flood water upstream and downstream from nearby I-40 in the Homer Wash Watershed in September 1994. The inspected embankments range from 1.8 to 2.4 m high and have slopes from about 2H:1V to nearly 1H:1V, or between 6 to 12 times steeper than what is proposed for the site flood protection berm. The steepest embankment, which deflected flood water toward an opening under I-40, appeared to be rip rapped with approximately 30 to 46 cm average size, graded, angular rock and placed to a thickness of 30 to 60 cm. This embankment training dike appears to remain very stable, despite its steep slope. The second, less steep, embankment, having an unprotected slope, except for desert vegetation, also appeared stable, despite its direct exposure to flood water in a major wash. No undercutting was evident along the toes of either embankment dike. The administrative record (LA Section 2410.6) states that the bridge and I-40 embankments were built to withstand erosion and flooding effects of the 100-year flood. **In the committee's judgement, the steep engineered embankments in the local area that they examined are capable of remaining**

stable for at least 25 years, even though they have been exposed to direct wash flow conditions since I-40 had been constructed.

No geotechnical stability analysis was performed on the 12H:1V external slope of the 1.5 m high flood protection berm for the site because of the low height and the relatively flat slope (LA Section 3100.3). The license application does not appear to present or address any plan for monitoring differential settlement in the 70-acre trench area.

Temporary Breakup Berms

In addition to the 1.5 m high flood protection berm, fifteen 0.3 m high flow "breakup berms" arranged in a chevron pattern, as described previously in this chapter, are proposed to be constructed west of and upgradient from the western-facing flood protection berm, to "slow and divert storm runoff to the north and south of the (LLRW) facility, and to ensure that flood flows remains subcritical near the permanent primary berm." The berms would be constructed during initial trench excavation from materials removed from the trenches, but not rip rapped (LA Sections 3310.2; 3200.5; 3440.1).

Earlier, it was pointed out that no clear hydrologic flood criteria or basis for designing the upslope temporary breakup berms was apparent in any reviewed LLRW documents.

At the end of a series of interrogations and responses on the breakup berms, U.S. Ecology's response to an interrogatory asking them to expand on the details of the breakup berms was:

> The small-flow breakup berms are constructed from the native material and are designed to withstand and divert the 100-year flood events. Once the primary berm is constructed and trench cover is complete, there will be no need for maintenance of the breakup berms (California DHS, 1990, Interrogatory No. 0216A Section 3100.1.1, 1990).

Therefore, although it is not apparent that the design of the temporary breakup berms was based on a consistent hydrologic criterion, i.e., 100-yr or PMF, the main purpose of the berms is to provide hydraulic roughness for slowing down the water velocity as it approaches the western flood protection berm at the LLRW site. The long-term integrity of the temporary breakup berms does not appear to be critical.

Site opponents, however, have raised the possibility of secondary effects resulting from postulated breaching of the berms. The Wilshire group postulated that "channelization of runoff by breaching of the berms, and integration of individual small channels into larger ones by capture induced by the berms" (Wilshire et al., 1994) could have one of two consequences: (1) runoff trapped by the N-S main flood control berm, ponding and potential infiltration and leakage into the trenches, and (2) undercutting of the rip-rap cover at the upslope corners of the main flood-control berm leading to its deterioration.

The Wilshire group cite, as anecdotal evidence for their concern about the stability of these breakup berms with respect to erosion and breaching, nearby graded road berms, an old

I-40 borrow pit area located at the northwest edge of the LLRW, and General Patton's base camp at the south end of Ward Valley (Wilshire et al., 1994). However, Wilshire later acknowledged during his August 30, 1994, presentation before the committee that these noted features may not have been engineered and constructed for the purpose of withstanding flooding or erosion forces.

The Wilshire group is correct in assuming that the small breakup berms will subsequently be eroded and breached over time. The small volume of sediment that may be eroded from the breakup berm will either wash past the flood protection berm in suspension (in the sheet wash) or deposit against the western upslope face of the rip-rapped flood protection berm. **It is the committee's view that the additional material will not adversely affect the stability or performance of the flood protection berm. In the committee's judgment, the erosion and breaching of the breakup berms over time are not a critical problem because the berms, in some form, will remain and offer some resistance to sheet flow, thus achieving their main objective although they were not designed as permanent structures.** As has been stated, their purpose is mainly redundant for *ensuring* subcritical flow at the downslope flood protection berm. **If breaching of these berms occurs and induces channelization and concentrates flow, however, the rip-rapped flood protection berm design is adequate, in the opinion of the committee, because there is still at least a 46 m safety margin distance between scouring at this location and the proposed trenches.**

In the committee's judgement, the flood protection berm and rip rap/filter system were effectively engineered to maximize the protection of the LLRW site from flooding and erosion, without apparently giving consideration to possible ponding or infiltration along the upstream edge of the flood protection berm toe. It is the committee's opinion that ponding along the relatively level upstream edge of the flood protection berm and potential infiltration and leakage into the adjacent trench zone are possible because of the highly permeable stone and gravel rip rap/filter layers along the exposed and subsurface parts of the flood protection berm slope. The possibility of slow vertical movement of ponded water into the underlying vadose zone or the possibility of perched saturated lenses developing in the unsaturated zone may enhance the potential for lateral movement toward the waste trench. Similarly, the Committee considers that the approximately three ha (seven acres) of exposed rip rap around all four sides of the flood protection berm will encourage directly-falling precipitation to infiltrate into the berm slope fill and possibly contribute recharge water to the trench zone over decades of exposure.

In summary, it is not clear what hydrologic design criterion was used for the small breakup berms, but this does not seem critical in terms of their intended function and lifespan. While the Wilshire group raises legitimate questions about possible channelization effects, such as scouring at the upstream corners of the downstream flood-protection berm, and possible ponding along the western edge of the flood-protection berm with seepage into the trench area, both of these concerns appear to be adequately addressed through effective flood-protection berm considerations.

The PMF Design

The use of Hydrometeorological (Hydromet) Report 49 data and procedures for estimating the Probable Maximum Precipitation (PMP) in the Colorado River and Great Basin Drainages (NOAA, 1977) is appropriate and reasonable as a basis for determining the PMF design flood.

The U.S. Army Corps of Engineers HEC-1 Flood Hydrograph Computer Package (1990); Soil Conservation Service (SCS) National Engineering Handbook- Section 4 (NEH-4) (1972); SCS TR-55 Urban Hydrology for Small Watersheds (U.S. SCS, 1986) for hydrologic modeling; and the Corps of Engineers HEC-2 Water Surface Profile Computer Package (U.S. Army, 1990) for hydraulic modeling are standard and reasonable methods for analyzing the assumed flood flows for flood protection and cover drainage facilities design and for predicting the potential flood impact of Homer Wash at the site.

Finally, the hydrologic analysis for both Homer Wash and local site PMF and 100-year flood peak computations apparently used SCS Weighted Curve Numbers, which were based on the assumption of *average* Antecedent Moisture Condition, or AMC II. While this design criterion assumption is reasonably conservative and meets the minimum regulations for the design of the perimeter flood protection berm, the use of the more conservative *saturated* soil-moisture condition, or AMC III, would have resulted in even higher peak flows for each of the assumed flood events. Application of AMC III for analyzing the Homer Wash PMP and 100-year floods, however, would not have any significant effect on the floodplain analysis since the site is already well outside and above the 100-year and PMF floodplains.

Historical Flood Experience

Because of limited development in Ward Valley, no surface-water flow has been monitored and no historical flow data are available for Homer Wash (LA Sections 2410.3.2, 2410.6). Regional streamflow records are, however, available for estimating the 100-year flood and a Regional Maximum Flood, for comparing with the computed 100-year flood and Probable Maximum Flood, respectively, for Homer Wash and at the local site.

Hydrologists at the U. S. Geological Survey have analyzed historical flood peaks for each of the six hydrologic regions of California, including the 129,500 km^2 South Lahontan-Colorado Desert (SL-CD) Region of Southeastern California, which contains the Ward Valley and Homer Wash (Waananen and Bue, 1977). Magnitude and frequency of floods were developed from regression analysis of flood peaks located at 43 stream gage stations in the SL-CD Region. For the SL-CD Region, the 100-year flood peak is defined as

$$Q_{100} = 1080 \ A^{0.71} \qquad\qquad \text{(Equation 7.1)}$$

where Q_{100} = 100-year peak flood flow (cubic feet per second [cfs]); A = Drainage area in square miles.

The 100-year flood regression equation is limited to a maximum watershed size of 65 km^2. The regression equation yields a peak flow of about 2790 cfs (79 m^3/s) for the 9.8 km^2 drainage area above the LLRW site (Figure 2.2). This represents about an order of magnitude higher peak flow than the 233 cfs (6.3 m^3/s) produced by HEC-1 computer simulation.

No 100-year flood comparison was made for Homer Wash, since this 127 km^2 watershed exceeded the watershed size limit for the regression equation.

Maximum observed peak flood flows from 883 sites in the conterminous United States, for drainage areas of less than 25,900 km^2, were analyzed by geographical regions by Crippen and Bue (1977). Envelope curves were computed that yield "reasonable limits" for estimating extreme flood potential for each region. Ward Valley and Homer Wash lie within Region 16 and the Regional Maximum Flood from the envelope curve produces about 29,000 cfs (820 m^3/s) and 190,000 cfs (5,400 m^3/s) for the local site drainage area (10 km^2) and Homer Wash (127 km^2), respectively. The local site Regional Maximum Flood peak is about three times the PMF peak of 10,270 cfs (290 m^3/s) from HEC-1 computer analysis and the 190,000 cfs (5,400 m^3/s) Homer Wash Regional peak is about an order of magnitude higher than the calculated PMF peak. No frequency or probability values were assigned to these Regional Maximum Floods.

The above comparisons of computed flood peaks, which were based on average soil moisture conditions and used in the LLRW facility design, to significantly higher regional flood peak potential, suggest that the conservative flood peak estimate used for the flood protection berm design may be exceeded and, therefore, should not be regarded as the outer limit of flood potential at the site.

GEOMORPHIC EVIDENCE OF EROSIONAL STABILITY

The Wilshire group (Wilshire et al., 1994) referred to older or late Pleistocene surfaces that:

> . . . are in the process of erosional degradation as seen by the degree of dissection and the disturbed pavements on remnants of the surfaces in medial and distal parts of the alluvial aprons.

In contrast U.S. Ecology, (LA Section 2310, p. 26), claims that:

> [t]he presence of surface and near surface relict and buried paleosols at the site indicates that the area is geomorphologically stable and has low sedimentation rates.

These statements seem to be important contrasts in understanding the stability of the surface and surficial deposits at the Ward Valley site. The geology of the alluvial fans and depositional processes provide a longer history pertaining to the surface stability and flooding potential for the site.

Drainage Incision and Alluvial Surfaces

The alluvial fan deposits along Ward Valley in the area of the proposed site have commonly coalesced into relatively uniformly sloping deposits or bajadas. The drainage channels, especially in the vicinity of the site, are very shallow and are interwoven into a braided network (Figure 2.3). The network of rills and shallow channels indicates little incising or downcutting within the channels. Connections in this network may change locally with time. All flow in these channels down the alluvial fan slope is ephemeral.

A single geomorphic surface covers much of the site area, and within the administrative record the thin alluvial deposits on this surface are designated Qf1. As described by Shlemon (Appendix 2310.A, Addendum C), Qf1 deposits show no soil development and are inferred to be geologically young and pedologically immature. The network of shallow channels is developed on this surface in the vicinity of the site.

Local areas are slightly higher than the surface of Qf1, and these areas are underlain by moderately developed relict soils or paleosols on which a surface designated Qf2 developed. From the trench descriptions, these Qf2 paleosols underlie much of the area, though they may be covered by very thin Qf1 deposits. Qf2 paleosols are developed similar to other paleosols in the region; from this Shlemon infers an age of 35,000 to 40,000 yr for the Qf2 paleosols.

A deeper paleosol (Qf3) exposed in some trenching at the site is also compared to other regionally developed paleosols. The age of Qf3 is estimated by Shlemon to be about 100,000 yr by comparison.

Evidence for Fan Deposition

The relationships between soil units and surficial deposits in the vicinity of the site provides evidence that there has been some slight regrading of the alluvial fan. This added a thin veneer of more recent sediment over much of the area. The smaller areas that are slightly higher and have older Qf2 at the surface are more vulnerable to erosion. They would be expected to be regraded if no baselevel changes or tilting occur. In general, however, the thin veneer of deposits indicates a relatively low net deposition at the site area.

The committee finds that surface deposits, paleosols, and a network of very shallow channels or rills are consistent with a surface in dynamic equilibrium for the past thousands to tens of thousands of years. Engineering analysis provides estimates of substantial potential channel erosion based on hypothesized large and infrequent rainfall and runoff events. **Nonetheless, in the committee's judgement, the surficial to near-surface deposits do not indicate such events under natural conditions for thousands of years in the past; from a geological perspective, the engineering estimates of channel erosion appear conservatively large.**

Homer Wash Erosion Potential

Homer Wash is an ephemeral stream draining Ward Valley from north to south. It has a low-sinuosity channel occupying approximately a mid-valley position. With a relatively high sediment flux from the alluvial fans and low runoff, Homer Wash has little potential to develop sinuosity or extensive floodplain deposits. Larger vegetation persists along the channel banks because of ephemeral runoff. Homer Wash and its equivalents in the past have apparently maintained much the same position, assuming the resistivity measurements partly reflect finer-grained deposits. Alluvial fan systems between these ranges will tend to maintain this position as well, as long as no uneven tilting occurs across the valley.

Conclusions from Geomorphic Evidence

1. The geomorphology of the Ward Valley site and immediate surroundings indicates dynamic equilibrium over tens of thousands of years. This reflects a certain balance between the climate and tectonics of the area. Changes in climate or relative base level would have to be significant and in the vicinity of the site to have an appreciable effect within a timeframe of less than thousands of years.

2. A significant baselevel change at the distance of Danby Dry Lake would require a long period of time before stream gradients would adjust and cause either erosion (because of a baselevel drop) or deposition (because of a baselevel rise) at the site.

3. If a significant fault scarp were to develop downslope of the site, it could cause considerable erosion and regrading across the site. The probability of such an event in this specific position appears to be low.

4. A significant increase in future precipitation could eventually change the gradient in Homer Wash and could result in some regrading across the site. However, as discussed above, the paleosol and other geomorphic evidence for long term stability over tens of thousands of years despite intervening major climate changes, as well as the stabilized, vegetated banks of the wash, suggest that any increase in precipitation would take some time to cause geomorphically significant changes to Homer Wash.

5. The site and immediate area show no discernible evidence of having undergone a change in their basic geomorphic stability.

CONCLUSIONS: ADEQUACY OF PROPOSED FLOOD PROTECTION SYSTEM DESIGN

Based on the foregoing discussion, observations, and analysis, the committee concludes the following concerning the potential for flooding and the engineered barriers and flood controls.

1. The hydrologic and hydraulic criteria, procedures and documentation used for analyzing the Homer Wash floodplain appear adequate. Because both the 100-year and Probable Maximum flood peak elevations on neighboring Homer Wash were estimated to be between 14.3 and 13.4 m below the LLRW facility site, potential flooding from Homer Wash is not considered to be a safety concern at the site.

2. The hydrologic and hydraulic criteria, procedures, and documentation used for designing the 1.5 m high flood protection berm appear to meet the minimum requirements for this type of facility. The proposed 12H:1V sloped flood protection berm and engineered rip-rap armoring system appear adequate to protect the berm from the long-term, desert surface runoff or sheet flow from a PMF event, which is postulated to occur during assumed average soil-moisture conditions. While this design criterion assumption is reasonably conservative, the design flood still could be exceeded, either assuming saturated conditions or considering regional historical flood data.

3. Erosion and breaching of the breakup berms will likely occur over a period of a few decades, but this is not critical to their main function and purpose in providing resistance to flow. Furthermore, any postulated channelization toward the LLRW site and any resulting scouring around the upstream corners of the flood protection berm appear to be adequately addressed by the above- and below-grade rip-rap design, especially because (a) at least a 46-m safety margin lies between scouring at this corner location and the LLRW material in the trenches and (b) waste is to be buried deeper than the estimated maximum scour depth during the design flood event.

4. The flood protection berm appears to have been effectively engineered to maximize the protection of the LLRW site from flooding, without giving consideration to possible ponding or infiltration along the upstream edge of the flood protection berm toe. Ponding along the relatively level upstream edge of the flood protection berm and resulting potential infiltration and leakage into the adjacent trench zone are possible because of the highly permeable stone and gravel rip rap/filter layers along the exposed and subsurface parts of the flood protection berm slope. The possibility of slow vertical movement of ponded flood water into the underlying unsaturated zone or the possibility of perched saturated lenses developing in the unsaturated zone may enhance the potential for lateral movement toward the waste trench.

5. The alluvial-fan geomorphology and shallow paleosols indicate dynamic equilibrium and little natural erosional potential. Ephemeral Homer Wash shows little evidence of floodplain development, consistent with engineering analyses placing the site well above the level of a hypothetical PMF or 100-year flood.

The above conclusions and opinions were reached, following a review of the methods, proposed design, and stated assumptions, and are considered to be within a degree of reasonable engineering and scientific certainty. No one can predict exactly *whether* or *when* the assumed design flood will occur or be exceeded by a catastrophic storm event and exactly *how* the facility will perform under either circumstance. Construction quality and commitment to long-term maintenance and performance monitoring are key factors in predicting how effectively a reasonably well-designed facility will meet its intended purpose.

Recommendations

The above analysis of flood protection facility design information and the breakup berms leads to the following recommendations for consideration if the site is developed for a LLRW facility.

1. While erosion and breaching of the breakup berms will likely occur over a period of a few decades, the berms will probably continue to provide flow resistance for several additional decades. If greater stability or longer-term performance is desired for these breakup berms, a more substantive design to minimize breach failure and possible channelization may be considered. One possible solution would be to use low, permeable rock dikes instead of local trench material. Such rock would provide desired roughness and be more stable over the lifetime of the facility, thus reducing the possibility of negative secondary effects such as the formation of small flow channels.

2. To address the possibility of ponding and infiltration, the committee recommends consideration of the use of an engineered channel (sloped to eliminate the possibility of ponding along the upstream edge of the flood protection berm and lined to reduce the possibility of infiltration into the underlying unsaturated zone), for conveying flood water around the west, north, and south sides and corners of the flood protection berm.

3. For additional infiltration protection, the committee recommends the use of an impermeable geosynthetic or alternative barrier under the flood protection berm's gravel filter layer to reduce both flood and non-flood rainfall event infiltration into the berm zone and lateral seepage toward the trench area.

4. The committee recommends testing the integrity of the proposed flood-protection berm for an assumed saturated antecedent soil-moisture condition (AMC III) when the PMP event is simulated. The designed flood-protection berm should be able to withstand this event, for at least the stillwater condition, with zero freeboard. In light of the relatively high Regional Maximum Flood peak potential compared to the computed design PMF peak, the committee recommends that the peer-review panel of hydrologic experts also assist DHS in reviewing extreme hydrologic event potential at the site and in recommending if any additional engineering design response is needed to defend the flood protection berm against such a rare flood event.

5. The committee recommends that more design and construction detail be provided for the rip-rapped chutes and outlets, which are proposed for conveying concentrated surface water drainage from the four cover swales while minimizing long-term scouring and infiltration.

6. In the committee's view, stability analysis should be conducted of 1.5-m high flood protection embankment slope to confirm stability under assumed seismic and toe-water conditions.

7. The committee recommends that a long-term monitoring plan be developed for detecting significant differential settlement of the trench-cover area and a response program for mitigating its potential negative effect(s) on surface drainage and floods. This plan also should include a comprehensive, operational and long-term flood and erosion facility monitoring and response program for identifying, repairing, or mitigating any stability, scouring, or sediment deposition problems which develop. The committee reemphasizes the Chapter 6 recommendation for an independent scientific oversight advisory panel of experts to review all monitoring data and the proposed response plan.

8. The committee recommends the development of a management plan for removing and properly disposing standing water from the open trenches during construction.

REFERENCES

Ad Hoc Interagency Committee on Dam Safety/Federal Coord. Council for Science, Engineering and Technology. Jun 1979. Federal Guidelines for Dam Safety. U. S. Govt. Printing Office, 041-001-00187-5.

Brandt, E. C. October 7, 1994. Summary submittal of the California Department of Health Services to National Academy of Sciences Committee to review specific scientific and technical issues related to the Ward Valley, California LLRW site. pp. 8, 32, 43.

California Department of Health Services (DHS), submitted by Roy F. Weston, Inc., Consultants. May 10, 1990. Interrogatories of concerns answered by U.S. Ecology, Inc. for License Application Sections Related to Flood Protection: Sections 3100, 3200 (No.s 0288A3200.2.1 and 0306A3200.5), and 3440 (No. 0457B3440.4).

California Department of Health Services (DHS). Aug 1993. Summary of comments from June-August 1991; comment period on final environmental impact report/statement (FEIR/S) and Department responses to State of California Indemnity Selection and Low-Level Radioactive Waste Facility. Sections 5.8 (Weather and Flash Flooding) and 5.11 (Erosion).

Chow, V. T. 1959. Open-channel hydraulics. New York: McGraw-Hill. Chapters 1- 8.

Crippen, J. R. and C. D. Bue. 1977. Maximum floodflows in the conterminous United States. U.S. Geological Survey, Water Supply Paper 1887:15, 52.

Federal Emergency Management Agency (FEMA). Sept 1985. Flood insurance study guidelines and specifications for study contractors. FEMA Report 37.

Franzini, D. L. Freyberg, and G. Tchobanoglous. 1992. Water-resources engineering. Ch. 7, 8.

License Application. 1989. U.S. Ecology, Inc. Administrative Record, Ward Valley Low-Level Radioactive Waste Disposal Facility, Sections 2310 (Geologic Site

Characterization), pp. 2310-26; 2410.3.1, 2410.3.2, 2410.6, 2410.7 (Surface Water Hydrology), pp. 2410-15, 2410-16, 2410-19, 2410-22, 2410-28, 2410-29, 2410-30; 3100.1.1, 3100.3 (Principal Design Features), pp. 3100-2, 3100-4, 3100-12; 3200.5 (Design Considerations for Normal and Abnormal Accident Conditions), pp. 3200-10; 3310.2 (Construction Methods and Features), pp. 3310-2, 3310-3; 3440.1, 3440.2 (Erosion and Flood Control System), pp. 3440-1, 3440-2, 3440-4; and 6310 (Surface Drainage and Erosion Protection).

Moser, D. A., 1993. An Application of Risk Analysis to the Economics Dam Safety, Engineering Foundation Conference Proceedings, Risk-Based Decision Making in Water Resources, eds. Y. Haimes and E. Stakhiv, sponsored by the National Science Foundation, Universities Council on Water Resources, U.S. Army Corps of Engineers, and American Society of Civil Engineers Task Committee on Risk-Based Decision Making. Santa Barbara, Calif. December, p.179.

National Research Council. 1983. Safety of Existing Dams: Evaluation and Improvement. Washington, D.C.: National Academy Press.

National Research Council. 1985. Safety of Dams: Flood and Earthquake Criteria. Washington, D.C.: National Academy Press.

Nuclear Regulatory Commission. Licensing requirements for land disposal of radioactive waste, 10 CFR 61, Subpart D, § 61.50 & 61.52. Jan 1, 1991.

San Bernardino County (SBC), California. 1986. Hydrology Manual. August.

Schreiber, D. L. Sept. 16, 1994. Detailed calculations and notes on the design for protection against flooding and erosion. Transmitted under cover letter to B.A. Tschantz.

U.S. Army Corps of Engineers. 1991. Hydraulic design of flood control channels. EM-1110-2-1601.

U.S. Army Corps of Engineers. Sept 1990. HEC-1, Flood hydrograph package and computer program 723-X6-L2010, CPD-1A, Version 4.0.

U.S. Army Corps of Engineers. Sept. 1990. HEC-2, Water Surface Profiles Manual and Computer Program, CPD-2A.

U.S. Army Corps of Engineers. 1976. Wave runup and wind setup on reservoir embankments. ETL-1110-2-221.

U.S. Army Corps of Engineers. 1971. Earth and rock-fill dams-general design and construction considerations. EM-1110-2-2300.

U.S. Department of Agriculture, Soil Conservation Service (SCS). June 1986. Technical Release 55. Urban hydrology for small watersheds (including TR-55 Interactive Micro-Computer Watershed Characterization Package, Version 1.11).

U.S. Department of Agriculture, Soil Conservation Service (SCS). 1972. SCS National Engineering Handbook, Section 4, Hydrology.

U.S. Department of Commerce, NOAA. Sept 1977. Probable maximum precipitation estimates, Colorado River and Great River Basins. Hydrometeorological Report No. 49.

U.S. Department of Commerce, NOAA. 1973. Precipitation - frequency atlas of the western United States. Vol. XI - California, NOAA Atlas 2.

U.S. Department of Commerce, U.S. Weather Bureau, prepared by D. M. Hershfield. Reprinted Jan., 1963. Technical Paper No. 40 (TP-40): Rainfall frequency atlas of the U.S.

U.S. Department of the Interior, Bureau of Reclamation. 1987. Design of small dams. Denver: U.S. Govt. Printing Office. 260 pp.

U.S. Ecology, Inc. 1990. Low-Level Radioactive Waste Management Facility Drawings CA-121-001,002,003,004,006,007,008,009, & 010.

U.S. Geological Survey. June 1977. Magnitude and frequency of floods in California. Water Resources Investigations 77-21.

U.S. Nuclear Regulatory Commission (NRC). Jan 1988. NUREG-1200 (Standard Review Plan 6.3.3, Sect. 4.3.2), Low-Level Waste Disposal Licensing Program.

U.S. Nuclear Regulatory Commission. Jan 1988. NUREG-1200, Standard Review Plan, Low-Level Waste Disposal Licensing Program, Rev. 1. Waananen, A. O. and J. R. Crippen. June 1977. Magnitude and frequency of floods in California. U.S. Geological Survey Water-Resources Investigations 77(21):6, 96.

Wilshire, H. G., K. A. Howard and D. M. Miller. Dec 1993. Description of earth-science concerns regarding the Ward Valley Low-Level Radioactive Waste Site Plan and evaluation.

Wilshire, H. G., H. G. Miller, and K. Howard. 1994. Ward Valley, Proposed Low-Level Radioactive Waste Site. A Report to the National Academy of Sciences.

8

THE DESERT TORTOISE

Issue 6
Potential Damaging Effects on the Desert Tortoise Habitat

THE WILSHIRE GROUP POSITION

Wilshire and others (1993) raised the concern that development of the Ward Valley site would have a serious effect upon the habitat of the desert tortoise (*Gopherus agassizii*) in an area with optimal geology and soils. They did not consider the mitigation plan established by U.S. Ecology, and supported by DHS, to be adequate to balance this loss.

THE DHS/U.S. ECOLOGY POSITION

The plan developed by U.S. Ecology included (1) a relocation plan for approximately 23-30 tortoises in the 36 ha site and other developed areas, a very small percentage of the total Ward Valley population; (2) fencing of about 10.5 km along Highway I-40, which bounds the site to the north, to protect the tortoise from further road kills; (3) use of various measures to reduce or prevent increases in predator populations; (4) speed control of low-level waste transport trucks on the access road; and (5) education of facility employees about desert tortoises. The components of this plan were endorsed by the U.S. Fish and Wildlife Services (USFWS) in its November 21, 1990 Biological Opinion (USFWS, 1990a) and were stated as the Reasonable and Prudent Measures to minimize incidental take. This opinion stated "that the proposed project is not likely to jeopardize the continued existence of the desert tortoise."

The Wilshire group also recently raised concern about potential spread of upper respiratory tract disease (URTD) if the tortoise population from the facility site were relocated into the Fenner Desert Wildlife Management Area (DWMA), one that presently is reported to have URTD in its western populations. Department of Health Services responded to this concern by questioning the accuracy of the health data used by the Wilshire group, indicating that the spread may be a result of handling tortoises in the health profile study plots.

THE COMMITTEE'S APPROACH

In our assessment of potential loss or alteration of habitat to evaluate the concerns of the Wilshire group, the committee included not only consideration of the loss of physical space and resources at the disposal site itself, but also possible changes in interactions between tortoises living in the vicinity of the facility and humans, native predators, and other tortoises.

History of Endangered Species Act Protection

Because the desert tortoise is protected as a threatened species by the Endangered Species Act, the listing history of this species is relevant to an assessment of impacts of the proposed facility. The desert tortoise occurs in southwestern North America on flats and bajadas with sandy-gravel soils (Luckenback, 1982). At least two distinct desert tortoise populations have been defined on the basis of genetic (Lamb et al, 1989) and morphometric criteria: the Mojave and Sonoran populations, and a possible third population in southern Sonora and Sinaloa, Mexico (USFWS, 1994a).

By the 1980's, desert tortoise populations had disappeared from parts of the western and northern Mojave Desert, and populations had declined in many other areas of the Mojave population's range (Jacobson, 1994; USFWS, 1994a). These declines led to the emergency listing of the entire Mojave desert tortoise population as federally endangered in 1989 (USFWS, 1989), and the final listing of the population as threatened on April 2, 1990 (USFWS, 1990b). The main reasons for this listing included habitat destruction, degradation, and fragmentation; removal or killing of tortoises by humans; increased predation on tortoises by native predators; disease; and failure of existing regulatory mechanisms to protect the desert tortoise and its habitat (USFWS, 1990b; 1994a). The Endangered Species Act mandates the designation of critical habitat and development of a recovery plan for federally listed species (16 U.S.C. § 1533 [1988 and Supp. 1990]). In response to this mandate, the U.S. Fish and Wildlife Service published the final critical habitat designation for the Mojave desert tortoise population on February 8, 1994 (USFWS, 1994b), and approved the final Desert Tortoise (Mojave Population) Recovery Plan (USFWS, 1994a) on June 28, 1994.

Critical habitat for the Mojave population was structured in accord with recommendations made in the Recovery Plan to include 14 Desert Wildlife Management Areas (DWMAs) in 6 Recovery Units. A fundamental goal of the Recovery Plan is to maintain the integrity of the distinct populations within each recovery unit, with DWMAs that contain desert tortoise habitat and viable tortoise populations identified in each unit. DWMAs are viewed as reserves, where human activities that negatively impact desert tortoises are restricted (Figure 8.1) (USFWS, 1994a). Ward Valley occupies parts of the Chemehuevi DWMA of the Northern Colorado Recovery Unit, and Fenner DWMA of the Eastern Recovery Unit, as designated in the Desert Tortoise (Mojave Population) Recovery Plan (USFWS, 1994a). The site for the proposed Ward Valley low-level radioactive waste (LLRW) disposal facility is located in the northern end of the Chemehuevi DWMA less than 2 km south of the Fenner DWMA, the two DWMAs being separated by I-40. The site was selected and characterization activities had begun before designation of the area as a desert tortoise critical habitat or identification of the DWMAs.

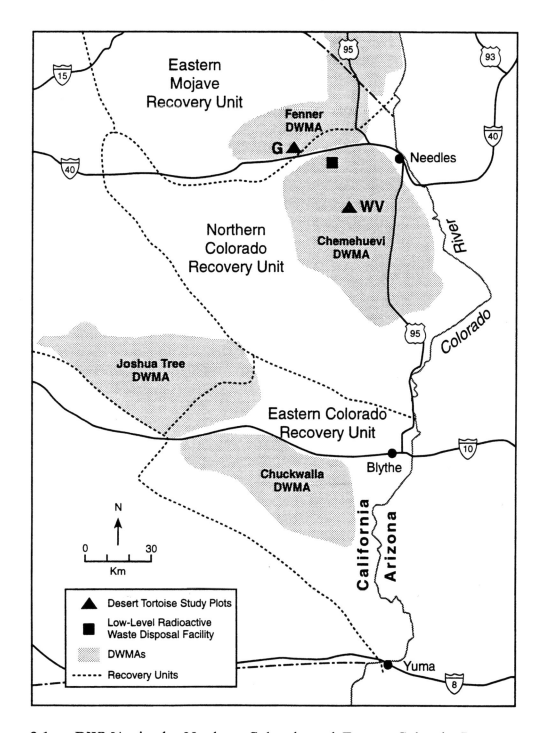

Figure 8.1 DWMAs in the Northern Colorado and Eastern Colorado Recovery Units identified by the Desert Tortoise (Mojave Population) Recovery Plan and approximate locations of the proposed low-level radioactive waste disposal site and the desert tortoise study plots (G, Goffs Health Plot, and WV, Upper Ward Valley Demographic Plot) (Modified from USFWS, 1994a, Berry 1994)

Potential Effects of the Facility on Desert Tortoise Habitat

The proposed disposal facility has four main potential effects on desert tortoise habitat.

Fragmentation of Desert Tortoise Habitat

First, the proposed facility will result in the direct loss of 36 ha and fragmentation of desert tortoise habitat within the DWMA considered by the Recovery Plan (USFWS, 1994a) to contain the largest and most robust of the remaining desert tortoise populations. This area includes 28 ha around the waste containment area itself that will be fenced to exclude tortoises, 3 ha to be occupied by buildings, and 5 ha that will be disturbed during construction of the flood control berms and other project-related facilities. Additional acreage will be lost through widening and upgrading the existing access road, which currently occupies less than 3 ha (USFWS, 1990b), and installation and use of the site-monitoring equipment in the 122 m wide buffer zone surrounding the fenced waste containment area (LA Section 3100.11). Based on reported estimates of tortoise home range size and density in this part of Ward Valley (Karl, 1989), the habitat loss for the facility itself would likely affect 18-20 tortoises, with an additional 5-10 tortoises affected by the access road upgrading (USFWS, 1990b). The fenced facility and the upgrading of the access road may also restrict movement of desert tortoises through part of Ward Valley. While the number of tortoises affected and the habitat area lost are small compared to the whole Ward Valley area, loss of habitat through fencing and road improvement must still be considered habitat fragmentation.

Increase in Predator Population

Second, the proposed facility may result in increases in the populations of animals that prey on young tortoises. Mortality rates of pre-reproductive tortoises are high, with predation by common ravens (*Corvus corax*) accounting for a high percentage of deaths (USFWS, 1994a). According to data collected as part of the U. S. Fish and Wildlife Service Breeding Bird Survey Program, raven populations in the Mojave Desert increased 15-fold from 1968 to 1988 (USFWS, 1994a). This population growth has been attributed to increased food supplies and perch and nest sites associated with human activity. An expansion of the local raven population could similarly accompany development of the Ward Valley disposal site.

Increased Contact with Humans

Third, interactions between tortoises and humans are likely to increase with construction and operation of the disposal facility. The 122 m wide buffer zone surrounding the fenced facility, which is for monitoring, equipment maneuvering, and remediation wells, will still be utilized by tortoises and will likely be a zone of increased contact with humans. The upgrading of the access road may attract additional traffic in the vicinity of the disposal facility (Karl, 1989). Detrimental effects of increased human activity include vehicular mortality along roadways (Nicholson, 1978), stress due to handling and other activities, and collection of tortoises for pets or other purposes (USFWS, 1994a).

Increased Competition Among Tortoises

Finally, the proposed project may result in increased competition among tortoises. This could occur in several different ways. The project plan currently calls for relocation of tortoises from the site to a nearby area that apparently has lower densities than the facility site (Karl, 1989). This could result in increased densities, and consequently increased competition, in the relocation area. If tortoises from the site itself were not relocated away from the site but merely excluded from their former territory by the fence around the waste containment area, this could similarly result in increased densities in the immediate vicinity of the facility. Even without increased tortoise density, levels of competition could increase if construction and operation of the waste disposal facility degraded the quality of the surrounding habitat and lowered its carrying capacity.

ASSESSMENT OF THE PLAN TO REMEDIATE POTENTIAL IMPACTS

The project proposal for the Ward Valley site comprises several approaches for mitigating adverse effects on the local desert tortoise population. These were initially based on the Bureau of Land Management (BLM) management plan for desert tortoise habitat (BLM, 1988), and include compensation for lost habitat, reduction of negative impacts on tortoises during facility construction and operation, and research to improve our understanding of desert tortoise ecology (Environmental Impact Report, 1993). These were also included as Reasonable and Prudent Measures in the USFWS November 1990 Biological Opinion.

Compensation for Habitat Loss and Fragmentation

The licensee intends to compensate for lost tortoise habitat with a two-part plan.

Fencing of I-40

The first step calls for fencing Interstate Highway I-40 and upgrading freeway underpasses to improve habitat currently supporting few tortoises and to facilitate movement throughout Ward Valley. The licensee maintains that fencing both sides of the freeway for

10.5 km where it crosses Ward Valley would enhance 1684 ha of tortoise habitat. This estimate is based on previous reports of reduced tortoise densities in the corridor adjacent to major highways (Nicholson, 1978; Karl, 1989), which has been attributed to increased roadkill and collecting. The estimate assumes the zone of strong freeway influence and low tortoise density is 0.8 km wide on either side of I-40.

The Committee's View of Fencing

This aspect of the mitigation plan has advantages and in all likelihood will reduce highway mortality and improve the potential for population increases in areas adjacent to the highway. However, the expected carrying capacity of the newly enhanced habitat along the highway remains unclear. Prior to selection of a specific site for the facility, the licensee performed a census of tortoises throughout section 34, the section that would have to be transferred to the state of California by the Department of the Interior (DOI) for construction of the facility. Based on those tortoise density data, the licensee selected one of the least dense parts of section 34 for the waste disposal facility (Karl, 1989). Facility construction and operation on the selected site would therefore affect fewer tortoises than on other possible sites in the immediate vicinity.

The fencing of I-40 and improvement of underpasses would likely facilitate north-south movement within the valley, because the freeway currently serves as a barrier to movement, with few tortoises using the existing underpasses (Karl, 1989). Although fencing would also probably reduce vehicular mortality, it is not clear whether traffic negatively affects tortoises in other ways that might limit occupation of the fenced habitat. For example, desert tortoises have been observed emerging from burrows in response to ground vibrations (USFWS, 1994a). Abnormal behavior, such as premature emergence from subterranean burrows or the apparent inability to detect and flee from predators, has been detected in other reptiles after experimental exposure to sounds of slightly lower intensities than those generated by off-highway vehicles (Brattstrom and Bondello, 1983). Even with a reduction in vehicular mortality from fencing, the area adjacent to the freeway may remain as suboptimal habitat for desert tortoises.

Relocation of Tortoises

The second part of the plan to compensate for lost habitat calls for relocating tortoises displaced during site construction into the protected habitat north of I-40 created through fencing along the highway. Approximately 23-30 tortoises will be moved as part of the relocation effort (USFWS, 1990b), including individuals removed from the fenced construction site and the area disturbed for berm construction. Other tortoises whose burrows are disturbed during construction will be relocated as determined at the time (USFWS, 1990a).

The Committee's View of Relocation

Although the relocation site will have geology and soils similar to the waste facility site, the committee sees several problems associated with the relocation plan. First, previous desert tortoise relocation studies have shown only limited success (e.g., Fusari et al., 1984; Burge et al., 1985; Baxter and Stewart, 1986). A combination of factors contributes to the generally low success rate of tortoise relocations: 1) the tendency of relocated tortoises to move long distances from the relocation site in either an erratic pattern or back towards their point of origin, 2) increased mortality among relocated individuals, 3) interference with the social structure of host tortoises, 4) increased stress on host and guest tortoises through competition and handling during relocation, and 5) introduction or enhanced transmission of disease (Fusari et al., 1984; Berry, 1986; Science Applications International Corporation, 1993; USFWS, 1994a).

Second, according to the guidelines for tortoise translocation in the Desert Tortoise (Mojave Population) Recovery Plan (USFWS, 1994a), displaced tortoises should not be released in DWMAs until relocation is much better understood. The proposed relocation, which would move tortoises from one DWMA (Chemehuevi) 2-3 km away to a different DWMA (Fenner) in a different Recovery Unit (Eastern Mojave), would be in opposition to this recommendation. The Recovery Plan encourages research on experimental relocation, but only outside DWMAs.

A third problem with the relocation plan is that the carrying capacity of the relocation habitat and the social structure of the host tortoise population are presently unknown. The Recovery Plan recommends that host tortoise populations be studied for at least two years before introducing tortoises into an area already inhabited by tortoises (USFWS, 1994a). Based on the abundance of tortoise signs tallied in strips along I-40 in Ward Valley, tortoises presently occupy the belt adjacent to the freeway, with the most dramatic depression of density north of I-40 being limited to the 0.40 km closest to the freeway (Karl, 1989). Without fencing to keep relocated tortoises in the strip of habitat along the freeway, guest tortoises are likely to wander into the home ranges of host individuals (either farther to the north or to the south on the other side of the freeway), where they may interfere with the social structure of hosts and compete with them for food and burrows (Berry, 1986).

Finally, the relocation plan could facilitate the transmission of disease from tortoises in the Fenner DWMA to individuals in the Chemehuevi DWMA. Desert tortoises in parts of the Mojave population are affected by an upper respiratory tract disease (URTD) and two shell diseases that have begun to receive intensive study only within the last five years. Disease has contributed to high mortality rates observed recently in the western Mojave Desert (USFWS, 1994a; Jacobson, 1994). Attention was focused on URTD by a severe outbreak of the disease in Desert Tortoise Natural Area in Kern County, CA, where 43 percent of 468 tortoises studied in 1989 showed signs of the disease, with a mortality rate of close to 25 percent (Jacobson et al., 1991). URTD may have been spread to wild tortoise populations by the illegal release of captive tortoises (Jacobson, 1993). A higher incidence of URTD has been reported near urban areas, which often have concentrations of captive desert tortoises, but the disease appears to be spreading (USFWS, 1994a). Habitat deterioration, poor nutrition, drought, and stress may all make tortoises more susceptible to URTD through immunosuppression (Jacobson et al., 1991).

In 1993, desert tortoises in the Goffs Health Profile study plot, approximately 13 km northwest of the proposed Ward Valley disposal site (Figure 8.1), tested seropositive for *Mycoplasma agassizii* when given ELISA tests (Berry, 1994). Schumacher et al. (1993) found ELISA tests to be an effective diagnostic tool for identifying the presence of *Mycoplasma agassizii*, a newly described mycoplasma that causes URTD in desert tortoises (Brown et al., 1994). Widespread shell abnormalities have also been observed in the Goffs Health Profile study plot (Berry, 1994), but the causes and effects of shell disease are not well understood (Jacobson et al., 1994). To date, URTD and shell disease have not been recorded at BLM's Upper Ward Valley study plot, located about 18 km south-southeast of the proposed disposal facility (Berry, 1994). The site licensors suggest that the presence of URTD at the Goffs study plot could represent localized transmission or aggravation of the disease caused by handling during research rather than a regional infection (Brandt, 1994). Although localized aggravation of the disease cannot be ruled out, introduction of desert tortoises from the proposed Ward Valley disposal site into an area closer to the Goffs study plot where disease has been detected is inadvisable. Particularly in view of the homing tendency of tortoises and their capability of moving long distances after relocation (as much as 7.3 km) (Berry, 1986), relocated tortoises could facilitate the spread of disease if they were exposed to ill tortoises near the Goffs plot, then moved back toward their original home range south of I-40.

Minimization of Predatory Population Growth

The mitigation plan for the Ward Valley LLRW site includes several measures aimed at minimizing the growth in raven populations that has been associated with increases in human activities in the Mojave Desert (Environmental Impact Report, 1993).

The top of the fence surrounding the waste containment area will be electrified to prevent the perching of ravens and other avian predators. Electrical transmission lines to the site will be buried, similarly to minimize the addition of perches for avian predators. Food waste and other garbage will be contained and roadkills along the access road will be collected daily to minimize the attraction of additional predators to the site. Bimonthly raven counts will be made and submitted annually to the U. S. Fish and Wildlife Service to monitor the size of local raven populations (USFWS 1990a).

Some increase in raven populations can be anticipated in the vicinity of the disposal facility, but the measures that will be implemented by US Ecology are likely to keep the increase small. Monitoring the raven population will allow a reevaluation and possible alteration of mitigation procedures if predator populations show a significant increase.

Measures to Reduce Impact of Tortoise/Human Interaction

Planned measures to reduce the impact of increased interaction between tortoises and humans include a speed limit and escort for transport vehicles on the access road, fencing of the 10.5-km segment of I-40 where it crosses Ward Valley, an education program for employees of the disposal facility, and monitoring of tortoise mortality.

The upgrading of the access road may attract increased traffic other than waste transport vehicles to the site (Karl, 1989). Other vehicles may not observe posted speed limits, which could result in occasional mortality of tortoises from roadkill. Increased traffic on the access road could also facilitate illegal collecting of tortoises. On the other hand, reduction in mortality from fencing I-40 will likely more than compensate for incidental mortality along the access road. The plans for monitoring tortoise mortality will permit the adoption of additional roadkill reduction measures, such as the installation of speed bumps, if the present plan proves to be inadequate for preventing roadkill mortality.

CONCLUSIONS AND RECOMMENDATIONS

The committee has two primary concerns about potential effects of the proposed facility on desert tortoise habitat: (1) limited habitat degradation and fragmentation associated with development of the facility, and (2) the unknown consequences of the relocation plan. The Recovery Plan and the critical habitat designation were both based on the fundamental principle of conservation biology that large, unfragmented habitat reserves are more effective in preventing extinction than small, fragmented ones (Simberloff and Abele, 1982; Wilcove et al., 1986). The Chemehuevi DWMA, which includes the proposed waste disposal facility, is the only DWMA in the Northern Colorado Recovery Unit and contains one of the largest and most robust desert tortoise populations in the Mojave Desert (USFWS, 1994a). It is currently the least fragmented and least disturbed DWMA and is considered to be a vital area for recovery of the desert tortoise (USFWS, 1994a). The Recovery Plan recommends the prohibition within DWMAs of surface disturbances that reduce the ability of the habitat to support tortoises (USFWS, 1994a). The acreage of habitat that will be permanently lost during construction and operation of the waste facility is small relative to the overall size of the Chemehuevi DWMA, but even small critical habitat losses may hinder recovery of the desert tortoise, as noted in the U. S. Fish and Wildlife Service's 1990 Biological Opinion pertaining to the proposed site (USFWS, 1990a, p. 18):

Development of such facilities within areas that may be deemed important for the recovery of the tortoise by the Recovery Plan or the Bureau's habitat management plans sets a biologically unsound precedent for future management actions.

- **The committee commends U.S. Ecology for its plans to decrease desert tortoise mortality in Ward Valley, but it recommends that the relocation plan be**

reevaluated in light of the Desert Tortoise (Mojave Population) Recovery Plan (USFWS, 1994a) and the paucity of data on successful tortoise relocations.

The Recovery Plan recommends that no displaced tortoises be released in DWMAs until relocation is better understood. Many unknowns remain about this particular relocation plan, in addition to the general uncertainties about relocation. These unknowns include the carrying capacity of the relocation habitat, the social structure and density of the host desert tortoise population, and the extent of infectious disease in the vicinity of the relocation area. The Recovery Plan recommends restricting relocations to areas outside DWMAs and, if the relocation area already supports tortoises, only after a 2-year study of the recipient habitat and population.

• If tortoise relocation remains as part of the mitigation measures for construction and operation of the Ward Valley facilities, the committee supports the recommendation of the U.S. Fish and Wildlife Service to establish a research program designed to study the effects of tortoise relocation and further recommends that relocation be made only outside DWMAs.

As part of the conditions in the Biological Opinion (USFWS 1990a), the U.S. Fish and Wildlife Service required establishment of a research program tied to relocation of the tortoises. If the research program is well designed, including intensive monitoring of tortoises relocated outside DWMAs, much could be learned about tortoise relocation and tortoise population interactions.

• As a possible alternative for the relocation plan, the committee suggests that consideration be again given to (1) evaluating impacts on the adjacent tortoise population of a plan that would exclude, but not relocate, resident individuals from all locations of facility construction and operation activities, or (2) consultation with the U.S. Fish and Wildlife Service about designating all individuals lost during construction as "incidental take".

Incidental take was addressed in the Biological Opinion, but in the opinion it referred to those tortoises lost after relocation of all collectable individuals. The committee feels that sacrifice of the tortoises on site or collection for adoption may ultimately have fewer adverse effects on the tortoise population in the vicinity of the facility than the proposed relocation.

• Finally, the committee recommends reinitiation of formal consultation with the U. S. Fish and Wildlife Service on the low-level radioactive waste disposal site at Ward Valley.

Reinitiation of consultation is required if one of the following conditions occurs: (1) the amount or extent of incidental take that was agreed upon is reached; (2) new information reveals effects of the agency action that may adversely affect listed species or critical habitat in a manner or to an extent not considered in the opinion; (3) the agency action is subsequently modified in a manner that causes an effect to a listed species or critical habitat that was not considered in the opinion, and (4) a new species is listed or critical habitat designated that may be affected by this action (50 CFR 402.16). The original Biological Opinion was made by the Service on November 21, 1990. Since that time, critical habitat that includes the proposed disposal site has been designated (USFWS, 1994b) and a recovery plan for the Mojave population of the desert tortoise has been developed and approved (USFWS, 1994a).

REFERENCES

Baxter, R. J., and G. R. Stewart. 1986. Report of continuing field work on the desert tortoise at Twentynine Palms Marine Corps Air Ground Combat Center, Spring 1985. Proceedings of the Desert Tortoise Council Symposium 1986.

Berry, K. H. 1986. Desert tortoise *(Gopherus agassizii)* relocation: Implications of social behavior and movements. Herpetologica 42:113-25.

Berry, K. 1994. Desert tortoise issues, in Ward Valley proposed low-level radioactive waste site: A report to the National Academy of Sciences: Presentations made to the National Academy of Sciences Review Panel, July 7 and 9 and August 30 to September 1, 1994, Needles, California, Wilshire, H., D. Miller, K. Howard, K. Berry, W. Bianchi, D. Cehrs, I. Friedman, D. Huntley, M. Liggett, and G. Smith, eds. pp. XV1-XV22.

Brandt, E. C. 1994. Summary submittal of the California Department of Health Services to National Academy of Sciences Committee to Review Specific Scientific and Technical Safety Issues Related to the Ward Valley, California, Low-level Radioactive Waste Site, October 7, 1994, Sacramento, California.

Brattstrom, B. H., and M. C. Bondello. 1983. Effects of off-road vehicle noise on desert vertebrates. Pp. 167-206, *in* Environmental effects of off-road vehicles: Impacts and management in arid regions, R. H. Webb and H. G. Wilshire, eds. Springer-Verlag, New York, New York.

Brown, M. B., I. M. Schumacher, P. A. Klein, K. Harris, T. Correll, and E. R. Jacobson. 1994. *Mycoplasma agassizii* causes upper respiratory tract disease in the desert tortoise. Infection and Immunity 62:4580-86.

Bureau of Land Management. 1988. Desert tortoise habitat management on the public lands: A rangewide plan. Washington, D.C.: U.S. Department of the Interior.

Burge, B. L., G. R. Stewart, J. E. Roberson, K. Kirtland, R. J. Baxter, and D. C. Pearson. 1985. Excavation of winter burrows and relocation of desert tortoises at the Twentynine Palms Marine Corps Air Ground Combat Center. Proceedings of the Desert Tortoise Council Symposium 1985:32-39.

Environmental Impact Report. 1993. Vol. 6 [14866-012], Section 2.1.6.1. in Administrative Record, California Low-Level Radioactive Waste Disposal Facility. Prepared for Department of Health Services/Bureau of Land Management by Dames & Moore.

Fusari, M., D. Beck, K. H. Berry, M. Coffeen, J. Diemer, and J. St. Amant. 1984. Problems with release of captive tortoises. Proceedings of the Desert Tortoise Council Symposium 1984:136-46.

Jacobson, E. R. 1993. Implications of infectious diseases for captive propagation and introduction programs of threatened/endangered reptiles. Journal of Zoo and Wildlife Medicine 24:245-55.

_____ 1994. Causes of mortality and diseases in tortoises: A review. Journal of Zoo and Wildlife Medicine 25:2-17.

Jacobson, E. R., J. M. Gaskin, M. B. Brown, R. K. Harris, C. H. Gardiner, J. L. LaPointe, H. P. Adams, and C. Reggiardo. 1991. Chronic upper respiratory tract disease of free-ranging desert tortoises *(Xerobates agassizii)*. Journal of Wildlife Diseases 27:296-316.

Jacobson, E. R., T. J. Wronski, J. Schumacher, C. Reggiardo, and K. H. Berry. 1994. Cutaneous dyskeratosis in free-ranging desert tortoises, *Gopherus agassizii*, in the Colorado Desert of southern California. Journal of Zoo and Wildlife Medicine 25:68-81.

Karl, A. E. 1989. Investigations of the desert tortoise at the California Department of Health Services proposed low-level radioactive waste facility site in Ward Valley, California, Report submitted to US Ecology and C. Robert Feldmeth, July 8, 1989.

Lamb, T., J. C. Avise, and J. W. Gibbons. 1989. Phylogeographic patterns in mitochondrial DNA of the desert tortoise *(Xerobates agassizi) [sic]*, and evolutionary relationships among the North American gopher tortoises. Evolution 43:76-87.

Luckenback, R. A. 1982. Ecology and management of the desert tortoise (Gopherus agassizii) in California. In North American tortoises: Conservation and ecology, R. B. Bury, ed. U.S. Department of the Interior Fish and Wildlife Service, Washington, D.C., Wildlife Research Report 12, pp. 1-37.

Nicholson, L. 1978. The effects of roads on desert tortoise populations. Proceedings of the Desert Tortoise Council Symposium 1978:127-29.

Schumacher, I. M., M. B. Brown, E. R. Jacobson, B. R. Collins, and P. A. Klein. 1993. Detection of antibodies to a pathogenic mycoplasma in desert tortoises *(Gopherus agassizii)* with upper respiratory tract disease. Journal of Clinical Microbiology 31:1454-60.

Science Applications International Corporation. 1993. American Honda tortoise relocation project: Final report. Science Applications International Corporation, Santa Barbara, California.

Simberloff, D., and L. G. Abele. 1982. Refuge design and island biogeographic theory: Effects of fragmentation. American Naturalist 120:41-50.

U.S. Fish and Wildlife Service. 1989. Endangered and threatened wildlife and plants; emergency determination of endangered status for the Mojave population of the desert tortoise. Federal Register 54(149):32326.

_____ 1990a. Biological opinion for the proposed low-level radioactive waste disposal facility, Ward Valley, San Bernardino County, California, (6840 CA-932.5) (1-6-90-F-41), November 21, 1990.

_____ 1990b. Endangered and threatened wildlife and plants; determination of threatened status for the Mojave population of the desert tortoise. Federal Register 55(63):12178-91.

_____1994a. Desert tortoise (Mojave population) recovery plan. U.S. Fish and Wildlife Service, Portland, Oregon.

_____1994b. Endangered and threatened wildlife and plants; determination of critical habitat for the Mojave population of the desert tortoise. Federal Register 59(26):5820-66.

Wilcove, D. S., C. H. McLellan, and A. P. Dobson. 1986. Habitat fragmentation in the temperate zone. Pp. 237-56 in Conservation biology: The science of scarcity and diversity, M. E. Soulé, ed. Sinauer Associates, Sunderland, Maryland.

Wilshire, H., K. Howard, and D. Miller. 1993. Memorandum to Secretary Bruce Babbitt, dated June 2. pp 3.

9

REVEGETATION

Issue 7

Potential Interference with Revegetation and Reestablishment of the Native Plant Community

THE WILSHIRE GROUP POSITION

The Wilshire group (Wilshire et al., 1993) raised the concern that long-term integrity of vegetation established on the covers of the waste repository trenches will be compromised because of "misconceptions about revegetation enhancements of the cover design." They emphasized that the raised aspect of the cover and its isolation from surface-water flow runon from the desert in the surrounding area would result in the lack of sufficient moisture to maintain vegetation on the cap. In support of their position, they cited publications that demonstrate the "dependency" of desert plants on soil recharge from surface flows (e.g., Schlesinger and Jones, 1984). Water erosion and infiltration through the cap was considered a consequence of the resulting absence of vegetation.

THE DHS/U.S. ECOLOGY POSITION

The DHS and U.S. Ecology point to their Revegetation Plan for the Ward Valley site in response to the Wilshire group's assessment. In the view of the site developers, this plan includes the potential for successful revegetation, not only through active revegetation but through natural revegetation from plant disseminules from the local area.

THE COMMITTEE'S APPROACH

Revegetation Plan

The committee reviewed the site revegetation plan, and other pertinent literature, and invited independent revegetation experts and experimenters to participate in the open meeting to share their knowledge and experience.

At the time of the review of the Ward Valley documents, US Ecology had not developed a comprehensive revegetation program. Their plans called for establishing a revegetation program with three phases: (1) procedures for transplanting cacti and yuccas during construction, (2) procedures for revegetation of caps of completed trenches during operations, and (3) procedures for restoration of the entire site after closure. Qualified biologists will be invited to participate on an ad hoc committee to help develop revegetation procedures and criteria for evaluating success of revegetation efforts. Selection of these biologists should be based on (1) fundamental knowledge of Mojave Desert ecosystems, for example, individuals who were involved with the International Biological Program (IBP) Mojave Desert validation site near Mercury, Nevada (see Wallace et al., 1980), (2) experience in restoring disturbed Mojave desert sites, and (3) familiarity with different biotic and abiotic components of desert ecosystems, for example, vegetation, vertebrates, invertebrates, and microorganisms. The revegetation program, which is planned to follow a set of guidelines established by US Ecology for their revegetation activities, such as use of native species and restoration of soil mycorrhizae, will be greatly strengthened if the committee of experts represents not only a broad spectrum of biological expertise, but also expertise in soil hydrology and ecology, and micrometeorology.

Although the revegetation program has yet to be developed, there are many individuals who have had extensive experience in revegetation of disturbed Mojave Desert locations. Vasek (1975 a,b) has used his knowledge of Mojave Desert ecosystems to restore disturbed pipeline corridors. Wallace (Wallace et al., 1980) was one of the leaders at the Mercury, Nevada IBP study site. Based on this multi-year effort, his group established guidelines and described problems that should be considered in desert restoration. Others have participated in these and similar efforts and could contribute considerable knowledge and real-world experience to any revegetation program at Ward Valley. The use of individuals with this level of desert restoration experience, and the development of a detailed revegetation program, is essential for this location. Although natural revegetation occurs on disturbed sites within the Mojave Desert, it may initially create a sparse vegetation cover that will be insufficient for its role as a soil stabilizer and transpiration agent (Vasek, 1979/80).

Consequences of Elevated Trench Cover

Elevation of the trench cover above the surrounding terrain will prevent runon and thus the upper end of the cover (caps) will receive only incident precipitation while the lower end will receive some runon from the upper end (limited because of the 518 m catchment length). A consequence of this moisture gradient will be a vegetation cover gradient, the upper end of the trench caps having sparser plant cover and possibly plants less robust than at the lower end.

The vegetation cover gradient should not cause a problem relative to soil erosion. The upper end of the trench cap will receive only rainfall and thus will not be impacted by surface flow erosion. On lower portions of the cover, and in the troughs between the trench caps, the increased runoff will be compensated by an increased vegetation cover. It is quite possible that soil erosion and water percolation may be no different between the upper and lower areas of the trench (Schlesinger et al., 1989). Planning and development of the revegetation program must take into account the potential occurrence of a natural moisture gradient and concomitant vegetation gradient, on the trench cap. Plantings of native plants should be designed to produce densities and cover equivalent to that expected in the high density areas on the lower end of the trench cap. This density, equivalent to the natural desert plant distribution, is expected to be maintained on the lower cap, while thinning will occur on the upper cap. However, climatic variables following active revegetation of the caps will control future densities, plant vigor and invasion by native species.

CONCLUSIONS AND RECOMMENDATIONS[1]

In the opinion of the committee, the guidelines presented as part of the revegetation plan have been put together with an understanding of desert plant ecology, and do not reveal any "misconceptions about revegetation enhancements", as charged by the Wilshire group (Wilshire et al., 1993). The emphasis on use of native species, establishment of appropriate desert soil conditions and mycorrhizae that will enhance moisture and nutrient uptake by transplanted and invasive plants, and protection of seedlings from insects and other animals is extremely important to the success of a revegetation program that will depend on both active transplanting and reinvasion by native plant propagules. Care should also be taken to reduce the number of non-native plants which potentially could occupy much of the newly prepared trench cap. The committee recommends that U.S. Ecology and its ad hoc committee of biological experts adhere closely to these guidelines in the design of the revegetation program.

It was not within this committee's charge to develop a revegetation program for this site. However, it is the committee's opinion that the revegetation program to be designed by US Ecology and the ad hoc committee of expert biologists, soil scientists, and micrometeorologists is likely to succeed if it takes into account (1) the guidelines for revegetation presented by US Ecology, (2) anticipated gradients in vegetation density and cover resulting from the moisture gradients on the trench caps, planning for higher density and letting natural conditions "thin" the community, and (3) the need for continual monitoring of the revegetation areas as part of the long-term monitoring program.

REFERENCES

Schlesinger, W. H., P. J. Fonteyn, and W.A. Reiners. 1989. Effects of overland flow on plant water relations, erosion and soil water percolation on a Mojave Desert landscape. Soil Science Society of America Journal 53:1567-1572.

Schlesinger, W. H., and C. S. Jones. 1984. The comparative importance of overland runoff and mean annual rainfall to shrub communities of the Mojave Desert. Botanical Gazette 145:116-124.

---

[1] M. Mifflin dissented from this conclusion. See Appendix F for his views on this issue.

Vasek, F. C. 1979/80. Early successional stages in Mojave Desert scrub vegetation. Israel Journal of Botany 28:133-148.

Vasek, F. C., H. B. Johnson, and G. B. Brum. 1975a. Effects of power transmission lines on vegetation of the Mojave Desert. Madrono 23:114-130.

Vasek, F. C., H. B. Johnson, and D. H. Eslinger. 1975b. Effects of pipeline construction on creosote bush scrub vegetation of the Mojave Desert. Madrono 23:1-64.

Wallace, A., E. M. Romney, and R. B. Hunter. 1980. The challenge of a desert: Revegetation of disturbed desert lands. Great Basin Naturalist Memoirs 4:214-218.

Wilshire, H., K. Howard, and D. Miller. 1993. Memorandum to Secretary Bruce Babbitt, dated June 2.

APPENDIX A

THE COMMITTEE'S CHARGE

COMMITTEE TO REVIEW SPECIFIC SCIENTIFIC AND TECHNICAL SAFETY ISSUES RELATED TO THE WARD VALLEY, CALIFORNIA, LOW-LEVEL RADIOACTIVE WASTE SITE

The committee will undertake an examination of the relevant data, reports, license application material, and other documents that address the following specific issues related to the Ward Valley, California, site:

1. Potential infiltration of the repository trenches by shallow subsurface water flow, including the possible presence of tritium in the deeper soils, and interpretation of C-14 ages.
2. Potential transfer of contaminants through the unsaturated zone to the ground water.
3. Potential for a hydrologic connection between the site and the Colorado River.
4. The absence of plans to monitor ground water or the unsaturated zone down-gradient from the site.
5. The potential for failure of proposed engineered flood-control devices.
6. Potential damaging effects on the desert tortoise habitat.
7. Potential interference with revegetation and reestablishment of the native vegetation.

The NRC/BRWM will arrange for the empanelment of a multidisciplinary scientific and technical review committee of about 18 experts from the fields of geology, geophysics, hydrology, geochemistry, civil engineering, and desert ecology. The committee will interview and interact with the scientists involved in the site study, state agency scientific staff, and others with expertise in desert hydrology, ecology, and geologic processes in two 3-day meetings. Two other 3-day meetings will be held after the relevant documents have been reviewed: the first for the committee to deliberate, discuss their findings, develop their conclusions, and prepare a report outline; the second, to complete the writing and integration of the report.

The objectives of the study are (1) to assess the adequacy of the site studies relative to the above enumerated issues and the validity of the conclusions concerning site performance that are the subject of debate, and (2) to determine if the enumerated concerns are valid, significant, and unresolved and, if so, to assess the potential impacts on site performance.

The committee will comment only on the scientific and technical issues. It will not evaluate the site nor be a party to any approval process.

The results of the study will be a report which will be reviewed and distributed in accordance with NAS/NRC procedures. DOI will receive 100 copies of the report; additional copies will be provided to BRWM committee members, state legislators, congressional representatives, public interest groups, environmental organizations and other parties in accordance with NRC policy. The report will be made available to all states and to the public without restriction.

APPENDIX B

CLASSIFICATION OF WASTES

(From the Code of Federal Regulations)

Title 10 - Energy
Chapter I - Nuclear Regulatory Commission
Part 61 - Licensing Requirements for Land Disposal
of Radioactive Waste
Subpart D - Technical Requirements for
Land Disposal Facilities

§ 61.55 Waste classification.

...(2) Classes of waste.

(i) Class A waste is waste that is usually segregated from other waste classes at the disposal site. The physical form and characteristics of Class A waste must meet the minimum requirements set forth in § 61.56(a). If Class A waste also meets the stability requirements set forth in § 61.56(b), it is not necessary to segregate the waste for disposal.

(ii) Class B waste is waste that must meet more rigorous requirements on waste form to ensure stability after disposal. The physical form and characteristics of Class B waste must meet both the minimum and stability requirements set forth in § 61.56.

(iii) Class C waste is waste that not only must meet more rigorous requirements on waste form to ensure stability but also requires additional measures at the disposal facility to protect against inadvertent intrusion. The physical form and characteristics of Class C waste must meet both the minimum and stability requirements set forth in § 61.56.

(iv) Waste that is not generally acceptable for near-surface disposal is waste for which form and disposal methods must be different, and in general more stringent, than those specified for Class C waste. In the absense of specific requirements in this part, such waste must be disposed of in a geologic repository as defined in part 60 of this chapter unless proposals for disposal of such waste in a disposal site licensed pursuant to this part are approved by the Commission.

(3) Classification determined by long-lived radionuclides. If radioactive waste contains only radionuclides listed in Table 1, classification shall be determined as follows:

(i) If the concentration does not exceed 0.1 times the value in Table 1, the waste is Class A.

(ii) If the concentration exceeds 0.1 times the value in Table 1 but does not exceed the value in Table 1, the waste is Class C.

(iii) If the concentration exceeds the value in Table 1, the waste is not generally acceptable for near-surface disposal.

(iv) For wastes containing mixtures of radionuclides listed in Table 1, the total concentration shall be determined by the sum of fractions rule described in paragraph (a)(7) of this section.

Table 1

Radionuclide	Concentration curies per cubic meter
C-14	8
C-14 in activated metal	80
Ni-59 in activated metal	220
Nb-94 in activated metal	0.2
Tc-99	3
I-129	0.08
Alpha emitting transuranic nuclides with half-life greater than five years	[1]100
Pu-241	[1]3,500
Cm-242	[1]20,000

[1]*Units are nanocuries per gram.*

Continued

(4) Classification determined by short-lived radionuclides. If radioactive waste does not contain any of the radionuclides listed in Table 1, classification shall be determined based on the concentrations shown in Table 2. However, as specified in paragraph (a)(6) of this section, if radioactive waste does not contain any nuclides listed in either Table 1 or 2, it is Class A.

(i) If the concentration does not exceed the value in Column 1, the waste is Class A.

(ii) If the concentration exceeds the value in Column 1, but does not exceed the value in Column 2, the waste is Class B.

(iii) If the concentration exceeds the value in Column 2, but does not exceed the value in Column 3, the waste is Class C.

(iv) If the concentration exceeds the value in Column 3, the waste is not generally acceptable for near-surface disposal.

(v) For wastes containing mixtures of the nuclides listed in Table 2, the total concentration shall be determined by the sum of fractions rule described in paragraph (a)(7) of this section.

Table 2

Radionuclide	Concentration, curies per cubic meter		
	Col. 1	Col. 2	Col. 3
Total of all nuclides with less than 5 year half life		700	(¹) (¹)
H-3	40	(¹)	(¹)
Co-60	700	(¹)	(¹)
N I-63	3.5	70	700
Ni-63 in activated metal		35	700 7000
Sr-90	0.04	150	7000
Cs-137		1	44 4600

1 There are no limits established for these radionuclides in Class B or C wastes. Practical considerations such as the effects of external radiation and internal heat generation on transportation, handling, and disposal will limit the concentrations for these wastes. These wastes shall be Class B unless the concentrations of other nuclides in Table 2 determine the waste to the Class C independent of these nuclides.

(5) Classification determined by both long- and short-lived radionuclides. If radioactive waste contains a mixture of radionuclides, some of which are listed in Table 1, and some of which are listed in Table 2, classification shall be determined as follows: (i) If the concentration of a nuclide listed in Table 1 does not exceed 0.1 times the value listed in Table 1, the class shall be that determined by the concentration of nuclides listed in Table 2.

(ii) If the concentration of a nuclide listed in Table 1 exceeds 0.1 times the value listed in Table 1 but does not exceed the value in Table 1, the waste shall be Class C, provided the concentration of nuclides listed in Table 2 does not exceed the value shown in Column 3 of Table 2.

(6) Classification of wastes with radionuclides other than those listed in Tables 1 and 2. If radioactive waste does not contain any nuclides listed in either Table 1 or 2, it is Class A.

(7) The sum of the fractions rule for mixtures of radionuclides. For determining classification for waste that contains a mixture of radionuclides, it is necessary to determine the sum of fractions by dividing each nuclide's concentration by the appropriate limit and adding the resulting values. The appropriate limits must all be taken from the same column of the same table. The sum of the fractions for the column must be less than 1.0 if the waste class is to be determined by that column. Example: A waste contains Sr-90 in a concentration of 50 Ci/m3. and Cs-137 in a concentration of 22 Ci/m3. Since the concentrations both exceed the values in Column 1, Table 2, they must be compared to Column 2 values. For Sr-90 fraction 50/150=0.33; for Cs-137 fraction, 22/44=0.5; the sum of the fractions=0.83. Since the sum is less than 1.0, the waste is Class B....

APPENDIX C

GENERAL TABLE OF COMMON CONVERSIONS

Appendix C - General Table of Common Conversions

In this report, units have been converted from English to metric where practical. Below is a general table of common conversions. With this conversion, some values may appear more precise than in the original document; for example, "less than 6 inches of rainfall" may now be written as "less than 15.2 cm of rainfall." One significant figure now appears as three. There is no easy remedy for this other than alerting the reader.

Linear Measure (Length)
1 centimeter (cm) = 0.3937 in
1 inch (in) = 2.540 cm
1 foot (ft) = 30.48 cm = 0.3048 m
1 mile (mi) = 1609 m = 1.609 km
1 meter (m) = 39.37 in = 3.281 ft
1 kilometer (km) = 0.6214 mi

Measure of Volume
1 cubic cm = 0.0610 in^3
1 cubic in = 16.39 cm^3
1 cubic ft = 28,317 cm^3 = 2.8x10^{-2} m^3
1 cubic m = 61,023 in^3 = 35.32 ft^3
1 acre-ft = 43,560 ft^3 = 1233 m^3
1 liter (l) = 1,000 cm^3 = 0.001 m^3
1 liter = 1.08 qt = 10.27 gal
1 gallon (gal) = 3.7 l = 3.7x10^{-3} m^3

Volumetric Flow Rates
1 ft^3/sec = 2.832x10^{-2} m^3/sec
1 m^3/sec = 35.32 ft^3/sec

Area Measure
1 sq cm = 0.155 in^2
1 sq in = 6.452 cm^2
1 sq ft = 929 cm^2 = .093 m^2
1 sq m = 1550 in^2 = 10.76 ft^2
1 hectare = 2.47 acres
1 acre = 43,560 ft^2
1 acre = 0.4047 hectare = 4.047x10^{-3} km^2
1 sq km = 0.386 mi^2
1 sq mi = 258.888 m^2 = 2.59 km^2

Weight
1 pound (lb) = 454 gm = 0.454 kg
1 kilogram = 2.20 lb

Unit Weight
1 lb/ft^3 = 16 kg/m^3
1 kg/m^3 = .06242 lb/ft^3
1000 kg/m^3 = 1 gr/cm^3
1 gr/cm^3 = 0.036 lb/in^3 = 62.4 lb/ft^3

APPENDIX D

GLOSSARY OF TERMS[1]

alluvium -

general term for unconsolidated material (clay, sand, silt, gravel) deposited by streams

bajada -

a broad alluvial slope extending from the base of a mountain range into a basin, formed by joining of a series of alluvial fans; occurs in arid and semi-arid conditions

base level -

theoretical lowest level toward which erosion of the Earth's surface constantly progresses; lowest level to which a river flows - locally could be into another river or lake; ultimately, into the ocean. Sea level is called the ultimate base level.

batholith -

a large mass of hardened molten rock with more than 100 km^2 of surface exposure with no known lower limit

berm -

a relatively narrow horizontal man-made shelf, ledge, or bench which breaks the continuity of a slope

breccia -

a coarse-grained rock composed of angular rock fragments held together by mineral cement

brittle deformation -

deformation in which rocks behave rigidly, by cracking and breaking, when a force or stress is applied

caliche -

secondary calcareous material found in layers on or near the surface in arid and semi-arid regions, composed of soluble calcium salts with varying amounts of gravel, silt, clay, and sand; a carbonate layer produced by soil-forming (pedogenic) processes

capillary force -

the action by which a fluid, such as water, is drawn up in small interstices as a result of surface tension of other droplets of water

clast -

a grain or fragment of rock, such as silt, sand, pebble or boulder

---

[1] Most definitions adapted from Bates, R. L. and J. A. Jackson, *eds.* 1980. Glossary of Geology. American Geological Institute, Second Edition. Falls Church, Virginia. pp. 751.

clastic -   pertaining to a rock or sediment composed of fragments of pre-existing rocks

compression -   a system of forces or stresses which result in the decrease in volume or shortening of the crust

cone of
depression -   a cone-shaped depression in the water table around a well which develops when water is withdrawn from the well

conglomerate -   a coarse-grained sedimentary rock composed of rock fragments larger than 2 mm in diameter and a fine grained matrix and/or cement of silt, sand, or calcium carbonate; the rock fragments usually are rounded, as pebbles, rather than angular as in a breccia

continental crust -   the crust which underlies the continents; mainly granite, a light colored, low density rock, high in silica and aluminum, low in iron and magnesium

continental margin-   edge of the continent which is between the shoreline and the abyssal ocean floor

continental shelf -   the part of the continental margin between the shoreline and the continental slope; the flat, gently sloping extension of the continent under the ocean

correlative -   belonging to the same stratigraphic position or level, generally implies strata of the same age, having formed at the same time in the geologic past

crust -   the outermost layer of the earth

crustal extension -   pulling apart of the outermost layer of the Earth's surface as a result of strain

crystalline rock -   a rock consisting wholly of relatively large mineral grains, refers generally to highly metamorphosed rocks

Darcy's Law -   a formula for calculating the rate of flow of fluids through a matrix of soil or porous rock on the assumption that flow is laminar and that inertia can be neglected

deep soil
water flux -   deep percolation of water in the unsaturated zone

deformation -   the process of folding, faulting, shearing, compression, or extension of rocks as a result of earth forces

deposition/
sedimentation -   the process of accumulation of loose rock material into layers or masses on the earth's surface above or below the sea

detachment fault -   a low-angle fault formed at the base of a horizontal or gently dipping body of rock; can be a thrust fault as a result of compression, or a normal fault as a result of extension.

dike -   an igneous intrusion which cuts across the layered structure of the surrounding rock; magma that fills a crack in the crust and hardens

dilution -   reduction in the concentration of a solution by addition of more solvent

dip -   an angle down from the horizontal that a sloping structural surface, e.g., a bed or a fault plane, makes

dispersion -   distribution of a substance outward in all directions

drainage system -   the network of surface streams, including a main stream and all of its tributaries, which drains a particular region

drawdown -   in a body of water, the difference between the water-surface elevation at a constriction, and what the elevation would be with no constriction

ductile deformation-   deformation in which rocks behave like a viscous substance

erosional
exhumation -   the uncovering or exposure by erosion of a surface, landscape, or feature that has been buried beneath other rocks

evaporite -   a non-clastic sedimentary rock composed of minerals produced from a saline solution, mainly the ocean or salt lake, due to evaporation, such as salt (sodium chloride) or gypsum

evapotranspiration -   loss of water from a land area through transpiration by plants and evaporation from the soil

extension -   a strain term signifying pulling apart the crust resulting in an increase in length

flux -   the rate of water movement through the unsaturated zone, regardless of direction

fold -   a curve or bend of a rock strata or other planar feature resulting from explosive volcanic activity that pulverizes rock as it blasts it out of the volcano

ground water -   subsurface water located in the saturated zone below the water table

hydraulic
conductivity -   a measure of the unsaturated zone's transmission of water

hydraulic
gradient -   in an aquifer, the rate of change of total head per unit of distance
                          in a given direction

infiltration -   the movement of water downward into soil or porous rock

island arc -   a chain of near shore islands rising from deep sea floor, produced by the
                          down-plunge of oceanic crust beneath a continental margin

isothermal -   maintaining a constant temperature during any process or procedure

lateral flow -   flow of water across a sloping surface or through a matrix because of gravity

lateral
subsurface flow -   water infiltrating the surface soil and moving laterally
                          in the subsurface following a shallow slope or gradient

magma -   molten rock, formed deep in the earth's crust, from which igneous rocks
                          solidify

magmatism -   the development and movement of magma

matric potential -   is a measure of how tightly water is held by the soil matrix

matrix -   the fine-grained material enclosing, or filling the interstices between, the larger
                          grains of a sediment or sedimentary rock

metamorphism -   the chemical, mineralogical, and/or structural change of rocks as a result of
                          change in temperature and/or pressure

mountain building -   the formation of mountains through the process of thrusting, folding, and
                          faulting of layers of the earth resulting from continental collisions. Deeper
                          layers of the crust also undergo metamorphism during this process.

oceanic crust -   the crust which underlies the ocean basins; mainly basalt, a dark, dense rock,
                          low in silica and aluminum, high in iron and magnesium

osmotic potential -   the energy required to remove dissolved salts from soil water

paleosol -   a buried soil horizon of the geologic past

perched water -   accumulated water trapped in the unsaturated zone by either some
                          impermeable layer or some structural feature

percolation -   the downward movement of water through small openings within a porous material, such as in the unsaturated zone

permeability -   the property or capacity of a porous rock, sediment or soil for transmitting a fluid; the interconnected pore spaces that allow movement of water from one place to another

piston flow -   the uniformly distributed downward movements of water in the unsaturated zone

playa -   a dry, vegetation-free flat area at the lowest part of an undrained desert basin; a dry lake bed in the desert

pluton -   an igneous intrusion; a large amorphous mass of magma formed deep within the earth, moved upward, and hardened before reaching the surface

potential -   refers to energy as function of position or of condition

precipitation -   the solidification of dissolved particles and settling out of solution by gravity

probable maximum precipitation (PMP)[2] -   the theoretical greatest depth of precipitation for a given duration that is physically possible over a given size storm area at a particular geographical location

probable maximum flood (PMF)[2] -   the most severe flood that is considered reasonably possible at a site as a result of meteorologic and hydrologic conditions

preferential flow -   movement of water downward through the unsaturated zone, along non-uniformly distributed pathways

Proterozoic -   the period of time before the Cambrian period, prior to macroscopic life

recharge -   the process of adding water to the saturated zone

regional uplift -   large-scale, long-term upward movement of an area of the crust

relief -   the vertical difference in elevation between hilltops or mountain summits and lowlands or valleys in a particular area. An area of high relief has great vertical variation and an area of low relief has little variation in elevation.

root zone -   area in the sediment in which living plant roots are found

---

[2] From Federal Emergency Management Agency. 1988. Glossary of Terms for Dam Safety.

saturation -   the point at which the interstices of a material, such as a rock, contain the maximum possible amount of water; all pores are filled with water

sill -   an igneous intrusion which parallels the planar structure of the surrounding rock; magma that forces its way between layers and then parallels the layered structure when it hardens.

soil-water
potential -   a measure of how tightly water is held by the soil matrix as result of capillary and other forces

sorption -   the taking up of a fluid or solution by a porous medium

standard deviation -   a quantification of the error range (±) of values about the average of a number of measurements

storage capacity -   the ability of a soil to hold water; the amount of water that can be held in the unsaturated zone

texture -   the general physical appearance or character of a rock; the pattern or interconnection of rock particles

tuff -   a general term for unconsolidated ash and fine fragments of rocks, resulting from explosive eruptions of volcanoes.

unconfined aquifer -   ground water that has a free water table, i.e. water not confined by pressure beneath impermeable rocks

unsaturated zone -   a subsurface zone including the soil that may contain water under pressure less than that of the atmosphere, including water held by capillary forces; it is the zone above the water table

vadose zone -   another term for unsaturated zone

volcanism -   the processes by which magma and its associated gases rise into the crust and are extruded onto the surface as lava flows and into theatmosphere

water table -   the surface between the saturated zone and the unsaturated zone, the uppermost part of ground water

water table divide -   a ridge or elevated zone in the water table from which ground water moves away in both directions

xeroriparian -   refers to species typically found in or along washes in arid environments

APPENDIX E

DISSENTING STATEMENT ON ISSUE 1

June Ann Oberdorfer
February 28, 1995

Of the seven issues raised by the Wilshire group, the first about transfer of contaminants to the water table is the most critical one from a human health and water resource perspective. The thick vadose [unsaturated] zone and dry climate are the first line of defense. Since the repository is designed as a passive system (no liners, no leachate collection, no monitoring or maintenance after the first 105 years after closure), it is imperative that our understanding of the rates of water movement through the vadose zone does not have large uncertainties. We need to have a reasonable assurance that any radionuclides which escape from the waste will not be able to migrate to the water table on time frames of concern.

The bulk of the reliable soil physics data and the chloride data point to minimal downward migration of soil water at the Ward Valley site. There are two sets of data, however, which prevent there being a reasonable assurance that there is minimal water flux at the site. The first of these is the detection of tritium in soil vapor samples at 30 m depth. If indeed there is tritium at that depth, then downward migration of soil water occurs at a much faster rate than any acknowledged in the license application and at a rate that would indicate the possibility of rapid transport of contaminants to the ground water.

The second data set indicates the apparent presence of a downward vertical hydraulic gradient in the ground water beneath the site. One reasonable explanation for this is that local ground-water recharge is occurring, with the flow occurring downwards away from this region to remove the added fluid mass.

United States Ecology, its consultants, and the Department of Health Services have treated both data sets as accurate without providing credible explanations for their occurrence. They have not questioned the validity of the data. The committee, on the other hand, has raised questions about the methods used in collecting these data. The validity of the data is unresolved and not resolvable without additional field work.

Since the data sets indicate much more rapid and more copious vadose zone water transport than acknowledged in the license application, they raise considerable concern about the suitability of the site for a LLRW repository. Without a resolution of this uncertainty by showing the absence of tritium at depth and the absence of a downward vertical gradient (or a defensible explanation for both phenomena), there cannot be reasonable assurance that water and contaminants will not be rapidly transmitted to the water table.

The report states that "site characterization data must be of sufficient quantity and quality to address the areal variability of percolation at a potential site..." The quantity and quality do not exist for Ward Valley to provide reasonable assurance that the site is suitable for long-term storage of radioactive wastes. Sampling of soil cores and analysis of soil moisture for tritium is not a difficult or lengthy task. Determination of the deviation from verticality of the wells is not one either. Satisfactory finding from these two tasks are important to dispel uncertainties about this site.

In addition, we need to be cautious about extrapolating from the field data to the engineered facility. The field data represent conditions and conclusions for the undisturbed site. Digging a large trench in the ground, filling it with containerized waste, and capping it with a raised cap greatly alter the natural system. There are many ways in which construction of the repository can modify infiltration and evapotranspiration. Subsidence of the waste can cause depressions in the cap surface where water could pond, particularly during times of heavy rainfall, and infiltration would then be greatly enhanced. Also during heavy rainfall periods, the troughs between the trenches will concentrate surface runoff into a small area (instead of its being spread over a very large area as in the natural system) where infiltration will be enhanced. Since the raised cap will have no surface water runon from upslope areas, much of the time the soil moisture will be reduced and hence the cap will not be able to support as dense a vegetative cover. During infrequent periods of heavy rainfall, more water may infiltrate than can be evapotranspired by the sparse vegetation before the water has time to percolate deep into the cap and then into the waste. Studies referenced in the report have found that unvegetated arid regions can have significant water fluxes and water fluxes could also be greater through a sparsely vegetated cap during periods of heavy rainfall. Only to the extent that the engineered facility mimics the natural system (no ponding, no concentrated channelization, sufficiently dense vegetative cover), can we argue that transfer of contaminants to the water table is unlikely.

APPENDIX F

DISSENTING STATEMENT ON ISSUES 1 AND 7

M. D. Mifflin
April 21, 1995

OVERVIEW

OPINION: Characterization studies at the Ward Valley Site failed to resolve the issue of deep percolation. Further, a history of both chemical and radioactive leachate at the water table at the Beatty facility, an analog of the Ward Valley Site, suggests inadequate documentation/understanding of the hydrologic processes in the thick vadose zones at both sites.

Arid-climate vadose zones are the least well-documented hydrologic systems of terrestial environments (and may prove to be the most deceptive as well). This Opinion disagrees with the Majority Opinion in its conclusions and recommendations for two of seven issues:

Issue 1:   The potential transfer of contaminants through the unsaturated zone to the ground water, and

Issue 7:   Potential interference of the raised trench design with revegetation and reestablishment of native plant community.

Issues 1 and 7 deal with the occurrence of water *below the root zone* that may percolate into the zone of emplaced waste and permit mobilization of contaminants through the formation of leachates. Issue 1 is related to vadose hydrologic processes that have been and are operating at the undisturbed site, and Issue 7 relates to processes that may operate when the site is used for waste disposal. Both issues relate to the role native vegetation in terms of infiltrated water and possible deeper percolation of water at the site. The Majority Opinion conclusions on Issues 1 and 7 are, in simple words, that the general knowledge and site databases indicate it is likely the vegetation root zones serve as efficient and effective barriers to deep percolation in the site area at the present time and that enough is known to reestablish vegetation to perform the same function on the trench covers.

The Majority Opinion discusses, in detail, the Ward Valley site databases with respect to strengths, weaknesses, and assumptions. Some additional observations are offered:

Saturated-Zone Water Samples and Analytical Utility: The pumped water samples at Ward Valley are vertically mixed samples due to the 14 meters of screened interval well design, and in *all* monitoring wells, the top of the screened interval is *5 meters or more below* the static water levels. These pumped water samples, therefore, do not represent the uppermost saturated-zone water samples, but rather mixed samples of primarily deeper flow zones in layered sediments of highly varied permeability. Thus, the saturated-zone water samples used to establish analyses for C-14, 0-18, deuterium, tritium, and water chemistry did not sample uppermost saturation and cannot give useful results from the perspective of site area recharge. Constituents of water samples representing vertical profiles in the upper part of the saturated zone allow recognition of stratified young water in the saturated zone, the presence or absence in turn would argue for or against significant recharge in the general area of the site. Increased concentration of modem carbon in a prolonged pumped sample from MW-WV-1 suggests that some stratification of younger water may be present at the site. There is too little known of the processes within the vadose zone and locations of recharge in arid terrain systems to establish confident interpretations as to "age" of water on the basis of pumped mixed layer samples. Varying mixtures of very young and very old water or, if only long-distance lateral flow is involved, only relatively old water, cannot normally be recognized. Further, the designs of the existing monitoring wells are not ideal for early recognition of leachate plumes in the saturated zone because of the stratified nature and varied permeability of the sediments in the saturated zone.

Uncertainty in Characterization Datasets: There are two Ward Valley site datasets that suggest that some deeper percolation and recharge may occur at the site: (1) the above background tritium values down to 30 meters in depth, and (2) the downward fluid potential gradient indicated by MW-WV-1 and MW-WV-2 water levels in the saturated zone. Soil moisture and chloride profiles indicate dry conditions and no percolation to below the root zone. Other Ward Valley site datasets also may be interpreted to permit some deep percolation by preferred or local paths, such as periodic water-level fluctuations, water-chemistry variations, high infiltration rates at the pit percolation test, high densities of normally deep-rooted plants, numerous small wash channels in the site area, a higher concentration of modern carbon in the longer-term pumped sample from MW-WV-1, the order of magnitude larger heat dissipation probe potentials than would be determined from the thermocouple psychrometers and soil moisture content data, and so forth. *However, all of these datasets are not definitive for the question being asked*: they are either extremely sparse, and/or often highly localized in some cases, ambiguous, or poor in quality or sampling design for the questions being asked, and therefore of uncertain utility with respect to confident documentation of the presence or absence of periodic pulses of infiltration (perhaps localized) and downward movement of water in the vadose zone.

Soil moisture monitoring studies in arid climates normally document little or no moisture percolating to more than 2 meters (Wierenga, et al., 1991; Estrella, et al., 1993; Gee, et al., 1994; Scanlon, 1994; Nichols, 1987; Fisher, 1992). However, evidence for deeper percolation and preferred pathways raise uncertainty (Allison and Hughes, 1983; Baumgardner and Scanlon, 1992; Elzeftawy and Mifflin, 1984; Gee and Hillel, 1988; Natir, et al., 1994; Scanlon, 1992; Steenhuis, et al., 1994). Direct moisture profile monitoring (potential and changing water content) and laboratory moisture content determinations in the vadose zone are very limited in both time and space. Other datasets, such as soil chloride profiles and tritium, are integrating the constituents of interest over various time periods. Most subsurface datasets do not integrate the sample over lateral areas beyond the borehole with the exception of gas phase (tritium) samples, and thus may not detect preferred pathways or relatively narrow and localized zones or corridors of matrix flow. Nearly all boreholes are located at the surface to avoid ephemeral surface washes (boreholes are typically located to protect the monitoring equipment or borehole operation from surface-water runoff events and associated losses). However, when channelized runoff occasionally occurs, it produces orders of magnitude greater infiltration per unit of time along narrow corridors as compared to incident precipitation, and many field experiments demonstrate that pulses of infiltrated water follow vertical pathways without lateral dispersion of more than one or two meters. Borehole-derived databases are biased to interfluve conditions for near-surface processes, and most sampled constituents are not suited to detect preferred pathway or highly localized percolation with the exception of pumped gas phase tritium. The significant tritium values at Ward Valley could be in error; however, they are the database of the vadose zone that has the best potential to document local deep percolation over a 40-year period, including areas beyond the boreholes.

The following give insight into the problems of short-term and highly localized vadose zone borehole monitoring databases. Elzeftawy and Mifflin (1984) applied soil physics monitoring techniques (thermocouple psychrometers) in Nevada to study recharge. Generally, no annual recharge seemed to occur in areas with less than 25 to 30 centimeters of mean annual precipitation. However, on the Jean-Goodsprings alluvial fan south of Las Vegas, Nevada (10-15 centimeters of mean annual precipitation) within 4 meters of a relatively well-developed wash channel, a 50-meter deep borehole drilled dry in the alluvium was equipped with a series of paired thermocouple psychrometers from one meter to depth. After reading several months of stabilized large negative potential readings at all depths, a saturation front passed to full depth (50 meters) immediately after a short-lived, high-intensity precipitation event of greater than 12.7 centimeters (the site storage gage topped over). Several hours of flow occurred in the wash. At another installation (thermocouples at 1.5 meters) in the Las Vegas area (Kyle Canyon fan, 20 centimeters of mean annual precipitation) only dry steady readings were observed for several years. However, nearby, a house basement in a disturbed area (natural vegetation removed) is 3 meters below existing grade but above and greater than 40 meters from a local wash. In the 25 years since construction, the normally bone-dry basement, in close contact with highly-developed calcrete layer (caliche), flooded once to a depth of 23 centimeters and drained in 48 hours. Infiltration apparently perched and flowed laterally on the calcrete layer to the basement area and up along the joints between the slab and walls toward the end of a 3-day series of heavy precipitation events (accumulated total of 23 centimeters).

Beatty Facility Experience: At the U.S. Ecology Beatty, Nevada, radioactive and hazardous waste disposal facility, the climate, vegetation zone, hydrogeology, and basin setting are similar to the Ward Valley site. The general design for the proposed Ward Valley disposal facility is closely based on the Beatty disposal facility, and similarites are pointed out throughout the license application documentation. Water balance and soil profile chloride studies generally support the notion of no active recharge at the Beatty site in an area immediately adjacent to the site during the period of studies (Prudic, 1994; Fischer, 1992; Nichols, 1987). The monitoring well analytical records, however, indicate that leachates from radioactive and chemical waste have reached the water table since 1982 (CRCPD D-5 Committee, 1994; Adams, 1990; Johnson, 1990). Because of the mixed evidence (for and against deep percolation) at both sites and the close similarities of the sites and activities, the experience at the Beatty disposal facility offers insight into the vadose zone hydrology of the Ward Valley site. However, the Committee had difficulty obtaining documentary records in a timely manner for the site and remains silent on the leachate evidence at the Beatty facility.

Modeled Performance: The Ward Valley site numerical modeling demonstrates anticipated benefits of the vadose zone retention capacity for any leachate that might develop. There are two unresolved questions, however, associated with the modeling exercise: (1) The existing moisture contents are unknown (assumed) at the Ward Valley site for the lower 600 feet of the vadose zone (an important weakness of the model), and (2) the same leachate retention modeled at the Ward Valley site should also be operating at the Beatty facility, where about 300 feet of vadose zone is available as moisture retention reservoir for percolating leachate. Ward Valley has, perhaps, about twice as much mean annual precipitation and double the thickness of the vadose zone as the Beatty facility. The modeled retention appears to not have occurred at the Beatty facility, as the first documented evidence for leachate in the saturated zone was in 1982, 20 years after disposal operations began, and apparent occasional breakthroughs of leachate continue to occur through 1993. Preferred pathways are a good possibility at the Beatty facility; large volumes of generated leachate are the other.

Deep Water Tables: There is a notion that deep water tables indicate little or no net recharge in arid terrains. A deep water table (thick vadose zone) may be related to a host of factors, the most important of which are a combination of limited availability of moisture for recharge *in the region* and the local *transmissive characteristics of the terrain* with respect to the overall configuration of the ground-water flow system (Mifflin, 1968). Vadose zone studies at Yucca Mountain (e.g. Fabryka-Martin, et al., 1993) highlight the above by demonstrating that an area with great depth to the regional water table (locally at over 400 meters) and similar climate to Ward Valley displays widespread evidence for some recharge (but the surface processes are still not fully documented). At Ward Valley, the deep water table <u>is</u> related to the general aridity of the region and flow-system configuration; absence of site-specific recharge should not be assumed on the basis of water-table depth.

Soil Chloride Profiles: Soil chloride profiles probably do not represent more than apparent histories of matrix transport/fate of incident precipitation at the borehole site. The age interpretations are directly dependent on assumed rates of chloride input (very, very uncertain in arid basin settings with extensive local salt sources, playas, and histories of marked climate and hydrologic changes). A series of chloride profiles taken from interfluve areas across the ephemeral washes and along the thalweg of washes would be of great interest. Such data do not exist. If the salt bulges were to consistently persist below washes, this line of evidence would be more convincing and useful.

Dry Drilling Technique Evidence: Other suggestive lines of evidence indicate deep percolation and associated recharge in basin environments of the Mojave Desert, but there is too little systematic data collection to establish how quantitatively important the basin recharge may be. In four undeveloped alluvial basins with known poor water quality in the Las Vegas area, dry drilling has been used in exploration for water supplies (Mifflin, unpublished field notes). Each basin explored by the drilling technique has produced suggestive evidence of localized recharge. Perching at depth (water produced during drilling above the level of stabilized static water levels) and uppermost (but of limited thickness) zones of low TDS water underlain by poorer-quality water has been noted. Such occurrences are interpreted as related to limited local recharge within the basins. The same dry-drilling approach (dual-wall reverse circulation rotary techniques available since the mid-1970's) would have produced better-controlled water samples and cuttings for the Ward Valley characterization effort.

Preferred Pathways: Desiccation fractures in silty and clayey sediments are well--documented in arid and semi-arid basins, including the Mojave and Great Basin Deserts. Such fractures are the best candidates for preferred pathways of flow in the vadose zone environments; they are documented to perform such a role in near-surface environments. Each alluvial basin fill has undergone at least five major climatic changes from pluvial climates to more arid climates during the Pleistocene, with attendant changes in positions of water tables, soil moisture conditions in the vadose zone, ground-water discharge and recharge relationships, and surface-water hydrology. Desiccation fractures and compaction faults attend the changing moisture contents in the fine-grained sediments. Earth fissures (Parker, 1963; Passmore, 1975; Patt and Maxey, 1978; Holzer, 1984; Mifflin, et al., 1991) are dramatic surface manifestations closely related to desiccation fractures, and most commonly are observed in areas where major declines in ground-water levels are taking place, usually in areas of heavy ground-water exploitation and attendant land subsidence; they also occur due to natural desiccation, often in playa areas. There is little question that, where earth fissures occur, very large volumes of ephemeral surface-water runoff pass to well below the root zone along preferred pathways. The hydrologic role the more deeply buried desiccation fractures may take is unknown.

Pulsed breakthroughs of contaminants at the Beatty facility can be explained without calling upon large volumes of leachate production if there are local, vertically extensive features such as desiccation fractures, and rapid downward flow of relatively small volumes of leachate along the fractures. Without preferred pathways, very large volumes of leachate seem necessary, and Issue 7 is elevated to a very important, poorly investigated and unresolved issue, with respect to operational practices and trench covers.

ISSUE 1

Opinion: **The site characterization failed to establish reasonable assurance of vadose zone hydrology of the site area. Two site-specific databases suggest deep percolation (the vadose zone showing a profile of significant levels of tritium and a saturated zone downward gradient), whereas other databases from six boreholes suggest little or no deep percolation at the borehole sites. The Beatty facility monitoring record and the significant tritium profile at the Ward Valley site combine to suggest the vadose-zone hydrology *is not understood at either site,* and that leachate may form and reach the water table much more rapidly than anticipated in the license application for Ward Valley.**

Recommendations: Beatty Facility studies are urged because they will likely yield important insight into the uncertain vadose-zone processes that permit leachate migration and better focused data collection at the Ward Valley site. (1) The Beatty facility waste containment experience warrants very careful, in-depth review and study. The leachate signals in the analytical record from the monitoring wells at the Beatty facility need to be independently reviewed from several perspectives: monitoring well designs, sampling methodologies, sample management, and laboratory analytical procedures, all of which have the potential to influence concentrations of indicators of leachates. (2) If radioactive and chemical leachates have traveled to the water table (as the records indicate and several agency reports conclude is the case), the independent review should expand to consider the relationships between facility design, operational practices, and leachate production. Trench closure techniques and current leachate production rates in open lined trenches are a critical part of the review. (3) If appropriate, the proposed Ward Valley site facility design, operation, monitoring, and closure plans should be reviewed and modified to minimize leachate production and downward migration.

Additional, carefully focused databases may help establish "reasonable assurance" and better background databases for monitoring system design requirements if the Ward Valley site goes forward as a waste-disposal facility. The following is believed the most cost-effective (both in time and funds) approach to developing the databases for resolution of the deep percolation issue: (1) Borehole deviation surveys in MW-WV-1 and MW-WV-2 will demonstrate if MW-WV-2 deviates from vertical sufficiently ($\geq 3°$) to produce the difference in water levels indicating a downward gradient in the saturated zone. (2) Additional boreholes for the vadose zone and uppermost saturated zone should be established by a dry drilling methodology that allows collection of moisture content cuttings and water samples *throughout* the vadose zone and several meters into the uppermost saturated zone. The dual wall reverse circulation rotary technique is cost-effective and useful for vertically controlled sampling in the 300 meter thick vadose zone and uppermost saturated zone. The methodology will produce cuttings in the vadose zone useful for the following analyses: Cl-36, soil moisture contents, water extractions for tritium (later gas extractions for tritium and C-14 are possible), and water samples from the uppermost saturated zone for tritium, C-14, environmental isotopes, and gross water chemistry. The boreholes can be finished for either a vadose zone and/or uppermost saturation monitoring. The air used for the drilling medium should be tagged with a tracer (SF-6). The methodology is flexible, and short intervals of coring can be adapted to the sampling techniques.

The actual drilling, sampling, and completion is not time-consuming. Between one and two weeks would be required for each borehole. Most analytical and interpretive work on samples should be complete within six months of sample collection. The most cost-effective completion approach is to finish the boreholes as uppermost saturated-zone monitoring wells.

ISSUE 7

OPINION: **The Wilshire Group concern about revegetation of raised trench covers, considering the width, configuration, and climate, is shared because of the absence of documentation of required infiltration adequate to *rapidly* reestablish and maintain a plant community equivalent to the high-density natural desert plant community, and a potential for the capillary barrier design to cause deep rooting to, and perhaps below, the capillary barrier horizon.**

The Wilshire Group raised concern about revegetation of the raised trench covers. They thought the raised trench cover design would not allow runon from adjacent areas and limited reestablishment of vegetation may allow infiltration through the cap and erosion due to the lack of stabilizing vegetation (Wilshire, et al., 1993). The Majority Opinion, however, generally concludes that there is no problem due to the current level of understanding of desert vegetation.

Discussion: The success or failure of long-term waste containment rests on two premises: (1) little or no infiltrated water reaching the emplaced waste to form leachate, and (2) what little leachate that may be generated is retained in the thick vadose zone due to the moisture retention capacity of the vadose zone. Therefore, timely revegetation to high densities is of paramount importance to assure little or no percolation of water to the emplaced waste.

A preliminary review of the trench cover and water balance literature suggests there is little or no documentation of the time and moisture required to reestablish the existing high-density vegetation and root-zone barrier on the raised trench covers in the Mojave Desert climate. The observations of Schlesinger, et al., (1989) and Schlesinger and James (1984) indicate surface-water runon is important, and the U.S. Ecology (1994) revegetated 25-year-old training dikes along I-10 (photographic evidence) may not be appropriate analogs of raised trench covers due to marked differences in configuration and surface-water availability. If revegetation to effective root barrier densities requires a decade or more, there is likely a very serious design problem that allows net accumulation of deeper percolation to accumulate on the capillary barrier at six meters. Gee, et al., (1994) have demonstrated up to 50% of precipitation may infiltrate in arid climates and percolate to greater depths under bare soil conditions. If two decades were to pass before an effective root barrier became fully established, and a net infiltration of 25% of mean annual precipitation occurred over the period, this amount equates to *4 to 5 meters of saturation above the capillary barrier* emplaced at 6 meters of depth. With the above scenario, the following would result: (1) roots would follow the moisture down due to the capillary fringe above perched saturation at the capillary barrier, and/or (2) flow breakthrough would occur in some parts of the capillary barrier or at least around the perimeter. The design purpose of the coarse capillary barrier layer within the trench cover is to prevent deeper percolation and limit the root depths. However, delayed revegetation might very well tend to encourage deep rooting, as well as eventually allow percolation of perching water into the waste horizon. Deep rooting phreatophytes, such as the mesquite of the local area, might become established if a near-surface capillary fringe were to develop.

Recommendations: (1) The raised trench cover and associated capillary barrier design warrant careful experimental evaluation at representative scales, where both the raised design trench cover and a trench cover at existing grade with natural runon are monitored and evaluated. Water balance studies with soil moisture and potential monitoring to at least the waste emplacement horizon is necessary. (2) Leachate production and rates of revegetation of trench covers should be part of the review study at the Beatty facility.

REFERENCES

Adams, S.R., 1990. Historical environmental monitoring report, U.S. Ecology Low-Level Radioactive Waste Disposal Facility, Beatty, Nevada, U.S. Ecology, September 27, 1990, 71 pp.

Allison, G.B., and M.W. Hughes, 1983. The use of natural tracers as indicators of soil-water movement in a temperate semi-arid region, Journal of Hydrology 602, pp. 157-173.

Baumgardner, R.W., Jr., and B. Scanlon, 1992. Surface fissures in the Hueco Bolson and adjacent basins, West Texas, University of Texas at Austin, Bureau of Economic Geology, Geological Circular 92(2), pp. 1-40.

Beven, K., 1991. Modeling preferential flow, *in* Gish, T.J., and A. Shirmohammadi (eds.), Preferential Flow, Proc. of the National Symposium, American Society of Agricultural Engineers, St. Joseph, MI, p. 77.

CRCPD E-5 Committee, October 1994. Environmental summary of the Beatty, Nevada Low-Level Radioactive Waste Disposal Site, Chapter 4, *in* Conference of Radiation Control Program Directors, Inc., Environmental Monitoring Report for Commercial Low-Level Radioactive Waste Disposal Sites, pp. 4-3 to 4-36.

Elzeftawy, A., and M.D. Mifflin, 1984. Vadose zone moisture migration in arid and semi-arid terrain, Symposium on Characterization and Monitoring of the Vadose Zone, Symposium Proceedings, National Water Well Association, Las Vegas, Nevada, December, 1983.

Estrella, R., S. Tyler, J. Chapman, and M. Miller, 1993. Area 5 site characterization project-report of hydraulic property analysis through August 1993, Desert Research Institute, Water Resources Center 45121, pp. 1-51.

Fabryka-Martin, J., M. Caffee, G. Nimz, J. Southon, S. Wightman, W. Murphy, M. Wickman, and P. Sharme, 1994. Distribution of chlorine-36 in the unsaturated zone at Yucca Mountain: an indicator of fast transport paths, Focus '93 Conference, Site Characterization and Model Validation, American Nuclear Society, Las Vegas, Nevada, pp. 56-58.

Fischer, J.M., 1992. Sediment properties and water movement through shallow unsaturated alluvium at an arid site for disposal of low-level radioactive waste near Beatty, Nye County, Nevada, U.S. Geological Survey, Water Resources Investigations Report 92(4032), pp. 1-48.

Gee, G.W. and D. Hillel, 1988. Ground-water recharge in arid regions: Review and critique of estimation methods, Journal of Hydrological Processes 2, pp. 255-266.

Gee, G.W., P.J. Wierenga, B.J. Andraski, M.H. Young, M.J. Fayer, and M.L. Rockhold, 1994. Variations in water balance and recharge potential at three western desert sites, Soil Science Society of America Journal 58, pp. 63-82.

Holzer, T.L., 1984. Ground failure induced by ground-water withdrawal from unconsolidated sediment, *in* Holzer, T. (ed.), Man-Induced Land Subsidence: Reviews in Engineering Geology, vol. VI, Geological Society of America, Boulder, Colorado, pp. 67-105.

Johnson, R.L., 1990. A review of organic contaminants in the unsaturated zone and groundwater zones at the Beatty, Nevada TSD Site, prepared for U.S. EPA, Region IX, Department of Environmental Science and Engineering, Oregon Graduate Institute, Beaverton, Oregon 97006, September 27, 1990, 8 pp. plus attachments.

Mifflin, M.D., Adenle, O.A., and Johnson, R.J., 1991. Earth fissures in Las Vegas Valley, 1990 inventory, Section C, *in* Bell, J.W. and J.G. Price (eds.), Subsidence in Las Vegas Valley, 1980-1991, Final Project Report, Nevada Bureau of Mines and Geology, pp. C1-C30.

Mifflin, M.D., 1968. Delineation of ground-water flow systems in Nevada, Desert Research Institute, CWRR, Technical Report Services H-W, No. 4, 111 pp.

Nativ, R., E. Adar, 0. Dahan, and M. Geyh, 1995. Water recharge and solute transport through the vadose zone of fractured chalk under desert conditions, Water Resource Research, 31:2, pp. 253-261.

Nichols, W.D., 1987. Geohydrology of the unsaturated zone at the burial site for low-level radioactive waste near Beatty, Nye County, Nevada, U.S. Geological Survey, Water Supply Paper 2312, p. 57 pp.

Parker, Gerald G., 1963. Piping, a geomorphic agent in landform development of the drylands, *in* International Association of Scientific Hydrology, Publication No. 65, Berkeley, California, pp. 103-113.

Passmore, Gary W., 1975. Subsidence-induced fissures in Las Vegas Valley, Nevada (M.S. Thesis), University of Nevada, Reno, Nevada, 105 pp.

Raft, R.O., and G.B. Maxey, 1978. Mapping of earth fissures in Las Vegas Valley, Nevada, University of Nevada, Desert Research Institute Project Report 51, 19 pp.

Scanlon, B.R., 1994. Water and heat fluxes in desert soils, 1. Field studies, Water Resources Research 30, pp. 709-719.

Scanlon, B.R., 1992. Evaluation of liquid and vapor water flow in desert soils based on chlorine 36 and tritium tracers and nonisothermal flow simulations, Water Resources Research 18, pp. 285-297.

Scanlon, B.R., 1992. Moisture and solute flux along preferred pathways characterized by fissured sediments in desert soils, Journal of Contaminant Hydrology 10, pp. 19-46.

Schlesinger, W.H., P.J. Fonteyn, and W.A. Reiners. 1989. Effects of overland flow on plant water relations, erosion and soil water percolation on a Mojave Desert landscape, Soil Science Society of American Journal 53, pp. 1567-1572.

Schlesinger, W.H., and C.S. Jones, 1984. The comparative importance of overland runoff and mean annual rainfall to shrub communities of the Mojave Desert, Botanical Gazette 145, pp. 116-124.

Steenhuis, T.S., J. Yves Pariange, and J.A. Aburime, 1994. Preferential flow in structured and sandy soils: Consequences for modeling and monitoring, in Wilson, L.G., L.G. Everett, and L.J. Cullen (eds.), Handbook of Vadose Zone Characterization and Monitoring, Lewis Publishers, Boca Raton, Florida, pp. 61-77.

U.S. Ecology, Inc., 1994. Handbook of supplemental information to the National Academy of Sciences, October 6, 1994.

Wierenga, P.J., R.G. Hills, and D.B. Hudson, 1991. The Las Cruces Trench Site: Characterization, experimental results, and one-dimensional flow predictions, Water Resources Research 27, pp. 2695-2705.

Wilshire, H., K. Howard, and D. Miller, 1993. Memorandum to Secretary Bruce Babbit, dated June 2, 1993, 3 pp.

REVEGETATION and WATER BALANCES
SUPPLEMENTAL REFERENCES

Berg, W.A., and P.L. Sims, 1984. Herbage yields and water-use efficiency on a loamy site as affected by tillage, mulch, and seeding treatments, J. Range Management 37(2), pp. 180-184.

Breshears, D.D., F.W. Whicker, and T.E. Hakonson, 1993. Orchestrating environmental research and assessment for remediation, Ecological Applications 3(4), pp. 590-594.

EPA, 1989. Final covers on hazardous waste landfills and surface impoundments, EPA/530-SW-89-047, July 1989.

EPA, 1985. Covers for uncontrolled hazardous waste sites, EPA/540/2-85/002, September 1985.

EPA, 1982. Evaluating cover systems for solid and hazardous waste, EPA#SW-867, GPO#055-000-00228-2.

EPA, 1972. Design and construction of covers for solid waste landfills, EPA-600/2-79-165, August 1979.

Essington, E.H., and E.M. Romney, 1986. Mobilization of 137Cs during rainfall simulation studies at the Nevada Test Site, *in* Lane, L.J. (ed.), Erosion on Rangelands: Emerging Technology and Data Base, Proc. Rainfall Simulation Workshop, January 14-15, 1985, Tucson, Arizona, ISBN: 0-9603692-4-4, Soc. Range Mgmt., Denver, Colorado, pp. 35-38.

Evanari, M., L. Shanan, N. Tadmor, and Y. Aharoni, 1961. Ancient agriculture in the Negev., Science 133(3457), pp. 979-996.

Foxx, T.S., G.D. Tierney, and J.M. Williams, 1984. Rooting depths of plants relative to biological and environmental factors, Los Alamos National Laboratory Report, LA-10254-MS.

Hakonson, T.E., 1986. Evaluation of geological materials to limit biological intrusion into low-level radioactive waste disposal sites, Los Alamos National Laboratory Report, LA-10286-MS.

Hakonson, T.E., L.J. Lane, and E.P. Springer, 1992. Biotic and abiotic processes, *in* Reith, C., and Thomson B.M. (eds.), Deserts as Dumps: The Disposal of Hazardous Materials in Arid Ecosystems, University of New Mexico Press, ISBN 0-8263-1297-7.

Hakonson, T.E., and L.J. Lane, 1992. The role of physical process in the transport of man-made radionuclides in arid ecosystems, *in* Harrison, R.M. (ed.), Biogeochemical Pathways of Artificial Radionuclides, John Wiley & Sons.

Hakonson, T.E., L.J. Lane, J.G. Steger, and G.L. DePoorter, 1982. Some interactive factors affecting trench cover integrity on low-level waste site, *in* Proc. Low Level Waste Disposal: Site Characterization and Monitoring, Arlington, Virginia, NUREG/CP-0028, CONF-820674, Vol. 2.

Hakonson, T.E., L.J. Lane, J.W. Nyhan, F.J. Bames, and G.L. DePoorter, 1990. Trench cover systems for manipulating water balance on low-level radioactive waste sites, *in* Bedinger, M.S. and P.R. Stevens (eds.), Safe Disposal of Radionuclides in Low-Level

Radioactive Waste Repository Sites, Low Level Radioactive Waste Disposal Workshop, USGS Cir. 1036, pp. 73-80.

Hakonson, T.E., L.J. Lane, G.R. Foster, and J.W. Nyhan, 1986. An overview of Los Alamos research on soil and water processes in arid and semi-arid ecosystems, in Lane, L.J. (ed.), Erosion on Rangelands: Emerging Technology and Data Base, Proc. of the Rainfall Simulator Workshop, January 14-15, 1985, Tucson, Arizona, Society for Range Management, Denver, Colorado, ISBN:0-960369264-4, pp. 7-10.

Hakonson, T.E., K.L. Manies, R.W. Warren, K.V. Bostick, G. Trujillo, J.S. Kent, and L.J. Lane, 1993. Migration barrier covers for radioactive and mixed waste landfills, in Proc. Second Environmental Restoration Technology Transfer Symposium, January 26-28, 1993, San Antonio Texas.

Hakonson, T.E., and J. W. Nyhan, 1980. Ecological relationships of plutonium in southwest ecosystems, in Hansen, W.C. (ed.), Transuranic Elements in the Environment, DOE/TIC-22800, U.S. Department of Energy, NTIS, Springfield, Virginia, pp. 403-419.

Hakonson, T.E., R.L. Waiters, and W.C. Hanson, 1981. The transport of plutonium in terrestrial ecosystems, Health Phys. 40, pp. 53-60.

Hakonson, T.E., G.C. White, E.S. Gladney, and M. Dreicer, 1980. The distribution of mercury, cesium-17, and plutonium in an intermittent stream at Los Alamos, J. Environ. Qual. 9, pp. 289-292.

Jacobs, D.G., J.S. Epler, and R.R. Rose, 1980. Identification of technical problems encountered in the shallow land burial of low-level radioactive wastes, Oak Ridge National Laboratory, SUB80/136/1, Oak Ridge, Tennessee.

Lane, L.J., 1984. Surface water management: a users guide to calculate a water balance using the CREAMS model, Los Alamos National Laboratory Report, LA-10177-M.

Lane, L.J., and F.J. Barnes, 1987. Water balance calculations in southwestern woodlands, in Proc. Pinyon-Juniper Conference, January 13-16, 1986, pp. 480-488.

Lane, L.J., and J.W. Nyhan, 1984. Water and contaminant movement: migration barriers, Los Alamos National Laboratory Report, LA-10242-MS.

Lane, L.J., E.M. Romney, and T.E. Hakonson, 1984. Water balance calculations and net production of perennial vegetation in the northern Mojave Desert, J. Range Management, 37(1), pp. 12-18.

Lane, L.J., J.R. Simanton, T.E. Hakonson, and E.M. Romney, 1987. Large-plot infiltration studies in desert and semiarid rangeland areas of the southwestern U.S.A., in Proc. International Conference on Infiltration Developments and Applications, University of Hawaii, January 6-8, 1987, pp. 365-376.

Lavin, F., T.N. Johnsen, and F.B. Gomm, 1981. Mulching, furrowing, and fallowing of forage plantings on Arizona Pinyon-Juniper ranges, J. Range Management 34(3), pp. 171-177.

McLendon, Terry, and E.F. Redente, 1991. Nitrogen and phosphorus effects on secondary succession dynamics on a semi-arid sagebrush site, Ecology 72(6), pp. 2010-2024.

Nyhan, J.W., G.L. DePoorter, B.J. Drennon, J.R. Simanton, and G.R. Foster, 1984. Erosion on earth covers used in shallow land burial at Los Alamos, New Mexico, Journal of Environmental Quality, 13, pp. 361-366.

Nyhan, J.W., and L.J. Lane, 1987. Rainfall simulator studies of earth covers used in shallow land burial at Los Alamos, New Mexico, *in* Lane, L.J. (ed.), Erosion on Rangelands: Emerging Technology and Data Base, Proc. of the Rainfall Simulator Workshop, Jan 14-15, 1985, Tucson, Arizona, Society for Range Management, Denver, Colorado, ISBN:0-9603692-4-4, pp. 39-42.

Nyhan, J.W., T.E. Hakonson, and B.J. Drennon, 1990. A water balance study of two landfill cover designs for semiarid regions, J. Environ. Qual. 19, pp. 281-288.

Nyhan, J.W., and L.J. Lane, 1986. Erosion control technology: a users guide to the use of the universal soil loss equation at waste burial sites, Los Alamos National Laboratory Report, LA-10262-M.

Romney, E.M., V.Q. Hale, A. Wallace, O.R. Lunt, J.D. Childress, H. Kaaz, G.V. Alexander, J.E. Kinnear, and T.L. Ackerman, 1973. Some characteristics of soil and perennial vegetation in northern Mojave Desert areas of the Nevada Test Site, University of California, Laboratory of Nuclear Medicine and Rad. Biology, Los Angeles, California, UCLA No. 12-916, UC-48 Biomedical and Environmental Research, TID-4500.

Screiber, H.A., and G.W. Frasier, 1978. Increasing rangeland forage production by water harvesting, J. Range Mangement 31(1), pp. 37-40.

Simanton, J.R., C.W. Johnson, J.W. Nyhan, and E.M. Romney, 1986. Rainfall simulation on rangeland erosion plots, *in* Erosion on Rangelands: Emerging Technology and Data Base, Proc. of the Rainfall Simulator Workshop, Tucson, Arizona, January 14-15, 1985, Society for Range Management, Denver, Colorado, pp. 11-17, 43-68.

Tierney, Gail D. and T.S. Foxx, 1987. Root lengths of plants on Los Alamos National Laboratory lands, Los Alamos National Laboratory, Los Alamos, New Mexico, LA-1 0865-MS, UC-48, 59 pp.

Tumer, R.M., 1982. Mojave desert scrub, Desert Plants 4(1-4), pp. 157-168.

U.S. Ecology, Inc., 1994. Handbook of supplemental information to the National Academy of Sciences, October 6, 1994.

Vasek, F.C., 1979/80. Early successional stages in Mojave Desert scrub vegetation, Israel Journal of Botany 28, pp. 133-148.

Vasek, F.C., H.B. Johnson, and G.B. Brum, 1975. Effects of power transmission lines on vegetation of the Mojave Desert, Madrono 23, pp. 114-130.

Vasek, F.C., H.B. Johnson, and D.H. Eslinger, 1975. Effects of pipeline construction on creosote bush scrub vegetation of the Mojave Desert, Madrono 23, pp. 1-64.

Wallace, A., E.M. Romney, and R.B. Hunter, 1980. The challenge of a desert: revegetation of disturbed desert lands, Great Basin Naturalist Memoirs 4, pp. 214-218.

APPENDIX G

BIOGRAPHICAL SKETCHES OF COMMITTEE MEMBERS

George A. Thompson, who chaired the committee, is professor emeritus, former chairman of the Department of Geophysics, former Dean of the School of Earth Sciences, at Stanford University. He was elected to the National Academy of Sciences in 1992. He is the recipient of awards from the Geological Society of America and the American Geophysical Union for his research and contributions to the study of the Basin and Range structure and tectonics, processes of extensional regimes, and deep crustal structure from seismic reflection and refraction measurements. He was an earth science consultant to the Advisory Committee on Reactor Safeguards, an independent committee that advises the Nuclear Regulatory Commission. He received a Ph.D. in geology from Stanford University.

Thure E. Cerling is professor of geochemistry at the University of Utah and for the last year has been visiting research scientist at the University of Lausanne, Switzerland. He has worked in several desert regions of the world. His major research interests are in soil geochemistry, the use of tritium as a hydrologic tracer, environmental geochemistry, dating of surface processes, and the geology of paleoanthropological sites. He earned a Ph.D. in geochemistry at the University of California (Berkeley).

G. Brent Dalrymple is dean of the College of Oceanic and Atmospheric Sciences at Oregon State University. He retired a geologist from the U.S. Geological Survey specializing in isotopic dating of geologic materials. He was elected to the National Academy of Sciences in 1993 and the American Academy of Arts and Sciences in 1992. He served as president of the American Geophysical Union and on the Executive Committee and Board of Governors of the American Institute of Physics. His interests and contributions to the field of isotopic dating include improvements to the instrumentation and methods of dating, recognition and dating of reversals of the earth's magnetic field, the history of Pacific plate motion, and lunar basin chronology.

Robert D. Hatcher, Jr. is Distinguished Scientist and professor of geology at the University of Tennessee and Oak Ridge National Laboratory. He was president of the Geological Society of America (GSA, 1993) and is president-elect of the American Geological Institute. He was given the first GSA Distinguished Service Award in 1988 after serving as editor of the GSA Bulletin. His primary research is directed toward the evolution of continental crust through the formation of mountain chains. He is also involved in studies of the relationships between fractures, hydrology, and containment migration, along with regional studies of fractures and intraplate seismicity. He received his Ph.D. in structural geology at the University of Tennessee.

Austin Long is professor in both the Departments of Geosciences, and Hydrology and Water Resources Research at the University of Arizona, as well as chief scientist at the University of Arizona's Laboratory of Isotope Geochemistry. His main fields of interest include Pleistocene geochronology, research in radiocarbon dating, and chemical equilibria and ground-water systems. He has done research on C-14 dating, transport of contaminants in the unsaturated zone, decision analysis methodology for assessing the performance of the proposed Yucca Mountain site, atmospheric C-13, and isotopic investigations of ground water in desert environments. He received his Ph.D. in geochemistry from the University of Arizona.

Martin D. Mifflin is president and senior hydrogeologist of Mifflin and Associates, Inc., a hydrologic consulting firm that has served as a consultant to the State of Nevada concerning the high level radioactive waste site at Yucca Mountain. He previously served as associate director and research professor at the Desert Research Institute Water Resource Center, University of Nevada System, for several years; Associate Professor of Geology at the University of Florida; and World Bank Resident Consultor to the National Water Plan of Mexico. Most of his studies and research have been related to hydrology in arid regions. He has a Ph.D. in hydrogeology from the University of Nevada.

June Ann Oberdorfer is professor of hydrology at San Jose State University. Her primary research is in evaluation of contaminated sites from hydrocarbon releases, numerical modeling of landfill contaminated sites, remedial action and aquifer test analyses, etc. She also consults for the Earth Science and Environmental Protection Divisions of Lawrence Livermore National Lab. She earned a Ph.D. in geology and geophysics from the University of Hawaii.

Kathleen C. Parker is an associate professor of geography at the University of Georgia. Most of her recent research centers on ecological relationships of cactus population, vegetation, and bird communities in the deserts of the southwestern U.S. She has been active on several committees and panels of the Association of American Geographers and has served on the panel on Conservation and Restoration Biology of the National Science Foundation. She received a Ph.D. in geography from the University of Wisconsin-Madison.

Duncan T. Patten is professor of botany at Arizona State University and Director of the Center for Environmental Studies. His research is in the ecology of montane and subalpine zones of the northern Rockies, ecology of desert plants, and heat and water flux within desert ecosystems. He served as scientific consultant to the National Science Foundation and consultant to the Public Land Review Commission on the environmental impact of open mining in the southwest. He has served on the panel on Environmental Biology and Ecological Sciences of the National Science Foundation and the Science Advisory Council of the Greater Yellowstone Coalition. He received a Ph.D. in botany/ecology from Duke University.

Dennis W. Powers is a private consulting geologist with broad experience and expertise that has been applied in site evaluation. He also specializes in sedimentology, evaporite geology, and environmental geology. He was involved in the site characterization of the Waste Isolation Pilot Plant (Carlsbad, NM) and served as supervisor of the Earth Sciences Division at Sandia National Laboratories. He received his Ph.D. in geology from Princeton University.

Stephen J. Reynolds is an associate professor at Arizona State University teaching tectonics, field mapping, and coupled tectonic/fluid processes. Prior to that he worked for the

Arizona Geological Survey, has mapped many areas of the southwestern Arizona desert, and published numerous papers on the geology and tectonics of the region. His research is in mid-Tertiary crustal extension in Arizona, core complexes, coupled tectonic/fluid processes. He received his Ph.D. in geology from the University of Arizona.

John B. Robertson is executive vice president and principal hydrogeologist of HydroGeoLogic, Inc., a hydrologic consulting firm specializing in Superfund site investigations and remediation. He retired from the U.S.Geological Survey. in 1984 as Chief of the Office of Hazardous Waste Hydrology. He has broad expertise and experience in ground-water hydrology related to quantitative analysis, treatment of contamination problems, chemical-physical-biological problems, siting criteria and assessment for low-level radioactive waste disposal. He chaired the Environmental Protection Agency Scientific Advisory Committee, National Center for Groundwater Research. He received his Geological Engineering Degree (M.S.) from the Colorado School of Mines.

Bridget R. Scanlon is currently research scientist with the Bureau of Economic Geology, University of Texas (Austin) and also teaches unsaturated zone hydrology at the university. Her research focuses on the use of soil physics, environmental tracers, and numerical simulations to quantify subsurface flow in arid regions. She served as a consultant to the Nuclear Waste Technical Review Board. She received her Ph.D. in geology at the University of Kentucky.

J. Leslie Smith is a professor of hydrology at the University of British Columbia. His research examines transport processes in fractured media, radionuclide transport in ground-water flow systems, thermal effects of ground-water flow, and the role of ground-water flow in geodynamic processes. He has received hydrology awards from the Geological Society of America and the American Geophysical Union (Meinzer and Macelwane). He consults for a variety of Canadian and U.S. agencies on projects related to site characterization, hazardous waste management, and peer reviews for work in these areas. He received his Ph.D. in geology from the University of British Columbia.

Bruce A. Tschantz is the R. M. Condra distinguished professor of civil and environmental engineering at the University of Tennessee. His research focuses on dam failure and safety analysis, hydrologic and river hydraulic modeling, and sediment transport modeling. He was chief of Federal Dam Safety for the Federal Emergency Management Agency (FEMA) and served as a consultant to the Office of Science and Technology Policy in the areas of dam safety and water policy; the Tennessee Departments of Conservation and Transportation; and the U.S. Geological Survey. He is a nationally-known engineering consultant and expert witness to industry, engineering firms, and attorneys in the areas of watershed hydrology and dam safety and has received national awards for both his teaching and research. He received his Sc.D in civil engineering from New Mexico State University.

Scott W. Tyler holds concurrent positions as associate research professor at the Desert Research Institute and associate professor at the University of Nevada (Reno), specializing in research in water and solute transport through unsaturated zones in arid environments. He has chaired Department of Agriculture research projects, and peer reviewed and edited several professional journals in hydrology, ground water, and soil science. He earned his Ph.D. in hydrology/hydrogeology from the University of Nevada (Reno).

Peter J. Wierenga is head of the Department of Soil and Water Science at the University of Arizona. His major research interests are processes of water flow and contaminant transport through saturated and unsaturated soils, movement of radionuclides, pesticides, heat, etc. and computer models of the processes. He is past chair of the Soil Physics Division of the Soil Science Society of America, and a former member of the board of the society. He has been awarded excellence in research awards, and consults for a variety of entities, including the Office of Technology Assessment, Environmental Protection Agency, Department of Energy, and several national labs. His Ph.D. is in soil science from the University of California (Davis).